Cognitive Technologies

Poramate Manoonpong

Neural Preprocessing and Control of Reactive Walking Machines

Towards Versatile Artificial
Perception–Action Systems

With 142 Figures and 3 Tables

 Springer

Author:

Poramate Manoonpong

Bernstein Center for Computational Neuroscience (BCCN)
Georg-August Universität Göttingen
Bunsenstrasse 10
(at Max Planck Institute for Dynamics and Self-Organization)
37073 Göttingen, Germany
poramate@nld.ds.mpg.de

Managing Editors:

Prof. Dov M. Gabbay
Augustus De Morgan Professor of Logic
Department of Computer Science, King's College London
Strand, London WC2R 2LS, UK

Prof. Dr. Jörg Siekmann
Forschungsbereich Deduktions- und Multiagentensysteme, DFKI
Stuhlsatzenweg 3, Geb. 43, 66123 Saarbrücken, Germany

ACM Computing Classification (1998): I.2.9, I.5.4, I.6.3, I.2.0

ISSN 1611-2482

ISBN   978-3-642-08835-3   e-ISBN   978-3-540-68803-7

Springer is a part of Springer Science+Business Media
springer.com

© Springer-Verlag Berlin Heidelberg 2007
Softcover reprint of the hardcover 1st edition 2007

Cover Design: KünkelLopka, Heidelberg

Printed on acid-free paper   45/3180/YL   5 4 3 2 1 0

To my family and
in loving memory of
my late grandfather,
Surin Leangsomboon

Foreword

Biologically inspired walking machines are fascinating objects to study, from the point of view of their mechatronical design as well as the realisation of control concepts. Research on this subject takes its place in a rapidly growing, highly interdisciplinary field, uniting contributions from areas as diverse as biology, biomechanics, material science, neuroscience, engineering, and computer science.

Nature has found fascinating solutions for the problem of legged locomotion, and the mechanisms generating the complex motion patterns performed by animals are still not very well understood. Natural movements provide the impression of elegance and smoothness, whereas the imitation of their artificial analogues still looks rather clumsy.

The diverse research on artificial legged locomotion mainly concentrated on the mechanical design and on pure movement control of these machines; i.e., in general these machines were unable to perceive their environment and react appropriately. Contributions developing embodied control techniques for sensor-driven behaviors are rare, and if considered, they deal only with one type of behavior; naturally, this is most often an obstacle avoidance behavior. There are only a few approaches devoted to the multimodal generation of several reactive behaviors.

This book presents a pioneering approach to tackle this challenging problem. Inspired by the obstacle avoidance and escape behaviors of cockroaches and scorpions, which here are understood as negative tropisms, and by the prey-capturing behavior of spiders, taken as a positive tropism, corresponding sensors and neural control modules are introduced in such a way that walking machines can sense and react to environmental stimuli in an animal-like fashion.

Besides obstacle avoidance, which is realised in indoor environments by using simple infrared distance sensors, other types of tropisms can be implemented by using diverse types of sensors. Especially, readers may find the introduction of hair sensors inspiring. These sensors are employed as contact sensors and at the same time serve as sound detectors, allowing for a "sound tropism".

One intriguing aspect of the presented neural technique is that it instantiates a very general design method. The neuromodules, manually constructed or developed with the help of evolutionary techniques, can serve as control structures for four-legged machines as well as for six- or eight-legged devices. The combination of a neural central pattern generator together with neural modules processing sensor inputs and modulating the output behavior points to an interesting opportunity for further developments. The simplicity of the utilised recurrent neural networks allows researchers to analyse and to understand their inherent dynamical properties. This makes the feasibility of an engineering approach to modular neural control even more convenient.

Heading towards autonomous walking devices, carrying all that is needed for an autonomous reactive behavior, i.e., energy supply, external sensors, and computer power, already makes the mechanical construction of these machines a difficult problem. The book provides some basic insights into biologically motivated mechanical constructions of four- and six-legged walking machines, which lead to robust platforms for robotic experiments.

Furthermore, the author demonstrates that using a modular neurodynamics approach to behavior control, which most efficiently acts in a sensorimotor loop, reduces the necessary computer power considerably. The multimodality of the described neural system furnishes these autonomous machines with convincing reactive behaviors.

The book provides a couple of ideas which can be taken up by students and researchers interested in the area of autonomous robots in general, and especially in the field of embodied intelligence. Autonomous walking machines are challenging systems because the coordination of many degrees of freedom has to be combined with a versatile set of external sensors. It should be noted that no proprioceptors, i.e., internal sensors like angle encoders, were used for generating walking patterns or behavior modulations. Together with this type of sensor, the mechatronical design methods and neural network techniques presented in this book will open up a new and wide domain of applications for autonomous walking devices. One really can congratulate the author for his achievements, for presenting this exciting research in general, and also for providing convincing practical examples in particular.

Sankt Augustin, Germany   Frank Pasemann
November 2006

Foreword

The motion and locomotion patterns of living things are complicated and have been neither easy to understand nor to imitate in action. However, it is well recognized by researchers around the world that such biological motion is naturally performed with the highest efficiency and effectiveness. Especially, biological reactive behaviors are considered as critical characteristics of animal survival in hostile environments.

Recognizing that, to date, there are quite a few prototypes of walking machines that can respond to environmental stimuli, the author has proposed a unique scheme of "Modular Neural Control" to be implemented in his walking machines. His scheme refers to a network containing multiple different modules and functionalities.

The simplicity of this network leads to an understanding of its inherent dynamic properties such as hysteresis profile and undesirable noise. This is then considered a major advantage, compared to the massive recurrent connections of traditional evolutionary algorithms. When applying this scheme to different types of walking machines, it requires less adaptation and changes of internal structure and parameters. This generic scheme enables walking machines to work in the real world, not just in a simulated environment. I think that the scheme will soon prove itself to be a pragmatic tool for the robotics design community.

An additional feature of this work is the "versatile artificial perception–action" system. An example of this versatility is the ability to perform more than one reactive behavior such as obstacle avoidance and sound tropism. It is our belief that an animal uses multiple reactive responses for survival in daily life.

While reading this book, I discovered that this research could also reveal a correlation between the complex walking behaviors of animals and their joint mechanism as well as the number of degrees of freedom. I have spent most of my life attempting to understand the relation between robotic structure and its function. It is clear from this book that the author has gone to a further step of a "designed" behavior of walking machines. I salute his achievement for doing exciting research in general and for getting practical results in particular.

Bangkok, Thailand   Djitt Laowattana
September 2006

Preface

The rationale behind this book is to investigate neural mechanisms underlying different reactive behaviors of biologically inspired walking machines. The systems presented here are formed in a way that they can react to real environmental stimuli (positive and negative tropism) using only sensor signals but no task-planning algorithm or memory capacities. On one hand, they can be used as a tool in order to properly understand embodied systems which, by definition here, are physical agents interacting with their environment. On the other hand, they can be represented as so-called artificial perception–action systems, which are inspired by an ethological study.

Most current physically embodied systems from the domain of biologically inspired walking machines have so far been limited to only one type of reactive behavior, although there are only few examples where different behaviors have been implemented in one machine at the same time. In general, these walking machines were solely designed for pure locomotion, i.e., without sensing environmental stimuli. This highlights that to date less attention has been paid to the walking machines which can interact with an environment.

Thus, in this book, biologically inspired walking machines with different reactive behaviors are presented. Inspired by obstacle avoidance and the escape behavior of scorpions and cockroaches, such behavior is implemented in the walking machines as a negative tropism. On the other hand, a sound-induced behavior called "sound tropism", in analogy to the prey capture behavior of spiders, is employed as a model of a positive tropism. The biological sensing systems which those animals use to trigger the described behaviors are investigated so that they can be reproduced in the abstract form with respect to their principal functionalities. In addition, the morphologies of a salamander and a cockroach capable of performing efficient locomotion are also taken into account for the leg and trunk designs of four- and six-legged walking machines, respectively.

Indeed, most of this book is aimed at explaining how to:

- Use a modular neural structure where the neural control unit can be coupled with the different neural preprocessing units to form the desired behavior controls. The neural structures are simple to understand and can be applied to control different types of walking machines.
- Minimize the complexity of the neural preprocessing and control unit by utilizing dynamic properties of small recurrent neural networks and applying by an evolutionary algorithm.
- Employ a sensor fusion technique to integrate the different behavior controllers in order to obtain an effective behavior fusion controller for activating the desired reactive behaviors with respect to environmental stimuli.
- Investigate morphologies of walking animals and their principle of locomotion control to benefit the design of the physical four- and six-legged walking machines.
- Achieve autonomous walking machines interacting with a real environment whereby the systems are challenged with unexpected real-world noise.

Acknowledgments

This book is a substantially revised version of my Ph.D. thesis, which was presented and submitted to the Department of Electrical Engineering and Computer Science at the University of Siegen.

The work described in this book could not have been done and the writing of this book would not have been possible without the help of numerous people. First and foremost, I would like to thank Prof. Dr.-Ing. Hubert Roth for accepting me as a doctoral candidate, for giving me the opportunity to pursue my Ph.D. studies at the University of Siegen and for supporting this work. I would like to especially thank Prof. Dr. rer. nat. Frank Pasemann from the Fraunhofer Institut für Autonome Intelligente Systeme for the continued guidance, for invaluable suggestions and discussions and for his availability at all times. I would also like to thank Dr. Djitt Laowattana and Dr. Siam Charoenseang from the Institute of Field Robotics, Thailand, for their encouragement and valuable advice throughout my educational career.

A special thank you goes to Dr. Jörn Fischer for his many valuable suggestions, ideas, and advice concerning software and hardware. I am very thankful to Manfred Hild for his suggestions regarding electronics. I am grateful to Dr. Bernhard Klaassen for his recommendations. I am also very thankful to Keyan Zahedi, Martin Hülse, Björn Mahn, and Steffen Wischmann for providing me with powerful simulation tools and for their useful advice.

Furthermore, I would like to thank my friends and colleagues Chayakorn Netramai, Arndt von Twickel, Fabio Ecke Bisogno, Matthias Hennig, Irene Markelic, Johannes Knabe, Sadachai Nittayarumphong, Azamat Shakhimardanov, Ralph Breithaupt, Winai Chonnaparamutt, and all other members of the INDY team for many helpful discussions, recommendations, inspiration, and the friendly atmosphere. I want to thank Wanaporn Techagaisiyavanit and Mark Rogers for being such faithful proofreaders.

Moreover, I would like to thank my parents and all members of my family for their encouragement and inspiration at all times during my studies and also for taking care of me. A special thanks to Siwaporn Bhongbhibhat for helping me to keep smiling and taking care of me. In addition, I am very grateful to all the people who have contributed the many useful materials to complete my work and who also offered many useful ideas and suggestions.

And last but not least, I would like to acknowledge the help and guidance provided by the Editor-in-Chief of the Springer Cognitive Technologies series, Prof. Dr. Jörg H. Siekmann.

Göttingen, Germany   Poramate Manoonpong
August 2006

Contents

1   **Introduction** 1
     1.1   Survey of Agent–Environment Interactions 1
     1.2   Aims and Objectives 8
     1.3   Organization of the Book 11

2   **Biologically Inspired Perception–Action Systems** 13
     2.1   Senses and Behavior of Animals 13
         2.1.1   Obstacle Avoidance Behavior 15
         2.1.2   Prey Capture Behavior 19
     2.2   Morphologies of Walking Animals 22
         2.2.1   A Salamander 23
         2.2.2   A Cockroach 24
     2.3   Locomotion Control of Walking Animals 26
     2.4   Conclusion 29

3   **Neural Concepts and Modeling** 31
     3.1   Neural Networks 31
         3.1.1   A Biological Neuron 32
         3.1.2   An Artificial Neuron 33
         3.1.3   Models of Artificial Neural Networks 36
     3.2   Discrete Dynamics of the Single Neuron 37
     3.3   Evolutionary Algorithm 41
     3.4   Conclusion 45

4   **Physical Sensors and Walking Machine Platforms** 47
     4.1   Physical Sensors 47
         4.1.1   An Artificial Auditory–Tactile Sensor 47
         4.1.2   A Stereo Auditory Sensor 49
         4.1.3   Antenna-like Sensors 51
     4.2   Walking Machine Platforms 55
         4.2.1   The Four-Legged Walking Machine AMOS-WD02 57

        4.2.2   The Six-Legged Walking Machine AMOS-WD06....... 61
    4.3   Conclusion 64

5   **Artificial Perception–Action Systems** 67
    5.1   Neural Preprocessing of Sensory Signals 67
        5.1.1   Auditory Signal Processing 68
        5.1.2   Preprocessing of a Tactile Signal 82
        5.1.3   Preprocessing of Antenna-like Sensor Data........... 87
    5.2   Neural Control of Walking Machines 89
        5.2.1   The Neural Oscillator Network 89
        5.2.2   The Velocity Regulating Network.................... 94
        5.2.3   The Modular Neural Controller 98
    5.3   Behavior Control 99
        5.3.1   The Obstacle Avoidance Controller 99
        5.3.2   The Sound Tropism Controller 103
        5.3.3   The Behavior Fusion Controller 104
    5.4   Conclusion 111

6   **Performance of Artificial Perception–Action Systems** 113
    6.1   Testing the Neural Preprocessing 113
        6.1.1   The Artificial Auditory–Tactile Sensor Data 113
        6.1.2   The Stereo Auditory Sensor Data 118
        6.1.3   The Antenna-like Sensor Data 123
    6.2   Implementation on the Walking Machines 126
        6.2.1   Obstacle Avoidance Behavior 128
        6.2.2   Sound Tropism.....................................135
        6.2.3   Behavior Fusion 142
    6.3   Conclusion 145

7   **Conclusions**...147
    7.1   Summary of Contributions...............................147
    7.2   Possible Future Work 149

A   **Description of the Reactive Walking Machines**.............151
    A.1   The AMOS-WD02.......................................151
    A.2   The AMOS-WD06.......................................152
    A.3   Mechanical Drawings of Servomotor Modules and the
          Walking Machines 154

B   **Symbols and Acronyms**...................................165

References..167

Index..183

1

Introduction

Research in the domain of biologically inspired walking machines has been ongoing for over 20 years [59, 166, 190, 199, 207]. Most of it has focused on the construction of such machines [34, 47, 216, 223], on a dynamic gait control [43, 117, 201] and on the generation of an advanced locomotion control [30, 56, 104, 120], for instance on rough terrain [5, 66, 102, 180, 192]. In general, these walking machines were solely designed for the purpose of motion without responding to environmental stimuli. However, from this research area, only a few works have presented physical walking machines reacting to an environmental stimulus using different approaches [6, 36, 72, 95]. On the one hand, this shows that less attention has been paid to walking machines performing reactive behaviors. On the other hand, such complex systems can serve as a methodology for the study of embodied systems consisting of sensors and actuators for explicit agent–environment interactions.

Thus, the work described in this book is focused on generating different reactive behaviors of physical walking machines. One is obstacle avoidance and escape behavior, comparable to scorpion and cockroach behavior (negative tropism), and the other mimics the prey capture behavior of spiders (positive tropism). In addition, the biological sensing systems used to trigger the described behaviors are also investigated so that they can be abstractly emulated in these reactive walking machines.

In the next section, the background of research in the area of agent–environment interactions is described, which is part of the motivation for this work, followed by the details of the approaches used in this work. The chapter concludes with an overview of the remainder of the book.

1.1 Survey of Agent–Environment Interactions

Attempts to create autonomous mobile robots that can interact with their environments or that can even adapt themselves into specific survival conditions have been ongoing for over 50 years [8, 41, 53, 75, 86, 136, 141, 143, 144, 157].

There are several reasons for this, which can be summarized as follows: first, such robotic systems can be used as models to test hypotheses regarding the information processing and control of the systems [69, 115, 146, 175]. Second, they can serve as a methodology for the study of embodied systems consisting of sensors and actuators for explicit agent–environment interactions [98, 99, 112, 135, 161]. Finally, they can simulate the interaction between biology and robotics through the fact that biologists can use robots as physical models of animals to address specific biological questions while roboticists can formulate intelligent behavior in robots by utilizing biological studies [63, 64, 173, 213, 214].

In 1953, W.G. Walter [208] presented an analog vehicle called "tortoise" (Fig. 1.1) consisting of two sensors, two actuators and two "nerve cells" realized as vacuum tubes. It was intended as a working model for the study of brain and behavior. As a result of his study, the tortoise vehicle could react to light stimulus (positive tropism), avoid obstacles (negative tropism) and even recharge its battery. The behavior was prioritized from lowest to highest order: seeking light, move to/from the light source, and avoid obstacles, respectively.

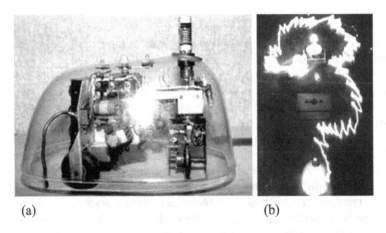

(a)   (b)

Fig. 1.1. (a) Walter's tortoise (photograph courtesy of A. Winfield, UWE Bristol). (b) The tortoise *Elsie* successfully avoids a stool and approaches the light (copyright of the Burden Neurological Institute, with permission)

Three decades later, psychologist V. Braitenberg [32] extended the principle of the analog circuit behavior of Walter's tortoise to a series of "Gedanken" experiments involving the design of a collection of vehicles. These systems responded to environmental stimuli through inhibitory and excitatory influences directly coupling the sensors to the motors. Braitenberg created varieties of vehicles including those imagined to exhibit fear, aggression and even love

(Fig. 1.2) which are still used as the basic principles to create complex behavior in robots even now.

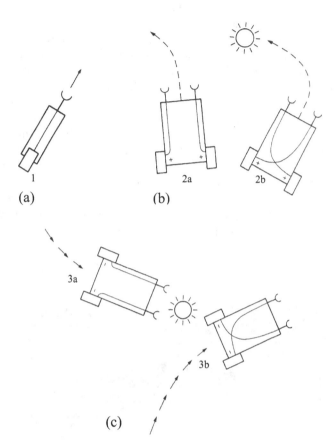

Fig. 1.2. Braitenberg vehicles. (**a**) Vehicle 1 consists of one sensor and one motor. Motion is always forward in the direction of the arrow and the speed is controlled by a sensor, except in the case of disturbances, e.g., slippage, rough terrain, friction. (**b**) Vehicle 2 consists of two sensors and two motors. Vehicle *2a* responds to light by turning away from a light source (exhibiting "fear"). Because the right sensor of the vehicle is closer to the source than the left one, it receives more stimulation, and thus the right motor turns faster than the left. On the other hand, vehicle *2b* turns toward the source (exhibits "aggression"). (**c**) Vehicle 3 is similar to vehicle 2 but now with inhibitory connections. Vehicle *3a* turns toward the light source and stops when it is close enough to the light source. It "loves" the light source, while vehicle *3b* turns away from the source, being an "explorer". (Reproduced with permission of V. Braitenberg [32])

One primitive and excellent example of a complex mobile robot (many degrees of freedom) that interacts with its environment appeared in Brooks' work [36, 38] in 1989. He designed a mechanism which controls a physical six-legged walking machine, *Ghengis* (Fig. 1.3), capable of walking over rough terrain and following a person passively sensed in the infrared spectrum. This mechanism was built from a completely distributed network with a total of 57 augmented finite state machines known as "subsumption architecture" [37, 39]. It is a method of decomposing one complex behavior into a set of *simple* behaviors, called layers, where more abstract behaviors are incrementally added on top of each other. This way, the lowest layers work as reflex mechanisms, e.g., avoid objects, while the higher layers control the main direction to be taken in order to achieve the overall tasks. Feedback is given mainly through the environment. This architecture is based on perception–action couplings with little internal processing. Having such relatively direct couplings from sensors to actuators in parallel leads to better real-time behavior because it makes time-consuming modeling operations and higher-level processes, e.g., task planning, unnecessary. This approach was the first concept toward so-called behavior-based robotics [10]. There are also other robots in the area of agent–environment interactions which have been built based on this architecture, e.g., *Herbert* [40], *Myrmix* [52], *Hannibal* and *Attila* [70, 71].

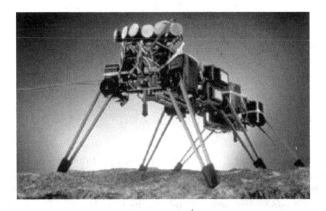

Fig. 1.3. The six-legged walking machine *Genghis*. It consists of pitch and roll inclinometers, two collision-sensitive antennas, six forward-looking passive pyroelectric infrared sensors and crude force measurement from the servo loop of each motor. (Photograph courtesy of R.A. Brooks)

In 1990, R.D. Beer et al. [22, 24] simulated the artificial insect (Fig. 1.4) inspired by a cockroach, and developed a neural model for behavior and locomotion controls observed in the natural insect. The simulation model was integrated with the antennas and mouth containing tactile and chemical sensors to perceive information from the environment; that is, it performs by wandering, edge following, seeking food and feeding food.

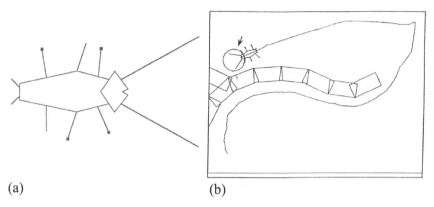

(a)   (b)

Fig. 1.4. (a) *Periplaneta computatrix*, the computer cockroach where the *black squares* indicate feet which are currently supporting the body. (b) The path of a simulated insect. It shows periods of wandering, edge following and feeding (*arrow*). (Reproduced with permission of R.D. Beer [22])

In 1994, Australian researchers A. Russell et al. [179] emulated ant behavior by creating robotic systems (Fig. 1.5) that are capable of both laying down and detecting chemical trails. These systems represent chemotaxis: detecting and orienting themselves along a chemical trail.

Fig. 1.5. Miniature robot equipped to follow chemical trails on the ground. (Photograph courtesy of A. Russell)

Around 2000, B. Webb et al. [212, 215] showed a wheeled robot that localizes sound based on close modeling of the auditory and neural system in the cricket (cricket phonotaxis). As a result, the robot can track a simulated male cricket song consisting of 20-ms bursts of 4.7-kHz sound. Continuously, such robot behavior was developed and transferred into an autonomous outdoor robot – $Whegs^{IM}$ ASP – three years afterwards [95]. The Whegs (Fig. 1.6) was able to localize and track the simulated cricket song in an outdoor environment. In fact, Webb and her colleagues intended to create these robotic systems in order to better understand biological systems and to test biologically relevant hypotheses.

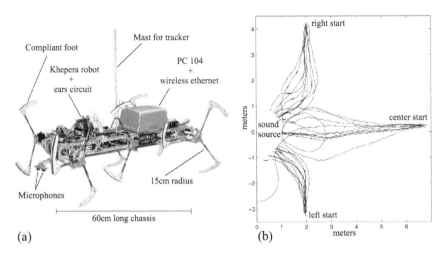

(a)   (b)

Fig. 1.6. (a) The Whegs. (b) Thirty sequential outdoor trials, recorded using the tracker, showing the robot approaching the sound source from different directions. (Reproduced with permission of A.D. Horchler [95])

The extension of the work of Webb was done by T. Chapman in 2001 [46]. He focused on the construction of a situated model of the orthopteran escape response (the escape response of crickets and cockroaches triggered by wind or touch stimulus). He demonstrated that a two-wheeled Khepera robot (Fig. 1.7) can respond to various environmental stimuli, e.g., air puff, touch, auditory and light, where the stimuli referred to a predatory strike. It performed antennal and wind-mediated escape behavior, where a sudden increase in the ambient sound or light was also taken into account.

In 2003, F. Pasemann et al. [155] presented the small recurrent neural network which was developed to control autonomous wheeled robots showing obstacle avoidance behavior and phototropism in different environments (Fig. 1.8). The robots were employed to test the controller and to learn about the recurrent neural structure of the controller.

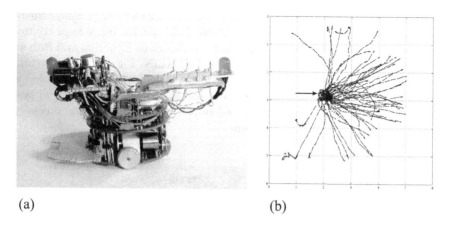

Fig. 1.7. (a) The robot model-mounted artificial hairs, antennas, ocelli and ear. (b) The combined set of wind-mediated escape run tracks, where the *arrow* indicates the stimulus. The robot was oriented in different directions relative to the stimulus. The *tracks* show the complete set of 48 escape run trials. (Reproduced with permission of T. Chapman [46])

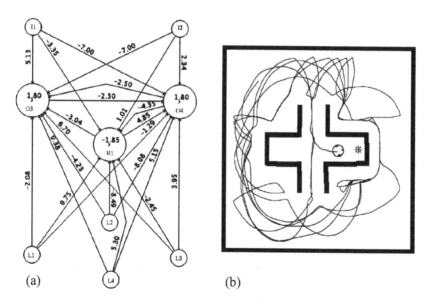

Fig. 1.8. (a) An evolved neural controller generating exploratory behavior with phototropism. (b) The simulated robot performing obstacle avoidance and phototropic behavior. (Reproduced with permission of F. Pasemann [155])

At the same time, H. Roth et al. [176, 177] introduced a new camera based on Photonic Mixer Device (PMD) technology with fuzzy logic control for obstacle avoidance detection of a robot called Mobile Experimental Robots for Locomotion and Intelligent Navigation (MERLIN, Fig. 1.9). The system was implemented and tested on a mobile robot, which resulted in the robot perceiving environmental information, e.g., obstacles, through its vision system. It can even recognize the detected object as a 3D image for precisely performing an obstacle avoidance behavior.

Fig. 1.9. MERLIN robots equipped with PMD cameras driving on a terrain with obstacles. (Reproduced with permission of H. Roth [177])

The above examples are robots in the domain of agent–environment interactions, a field which is growing rapidly. The most comprehensive discussion can be found in the following references: R.C. Arkin (1998) [10], J. Ayers et al. (2002) [11] and G.A. Bekey (2005) [26].

1.2 Aims and Objectives

The brief history of the research presented above shows that the principle of creating agent–environment interactions combines various fields of study, e.g., the investigation of the robotic behavior control and the understanding of how a biological system works. It is also the basis for the creation of a so-called Autonomous Intelligent System, which is an active area of research and a highly challenging field. Thus, the work described here continues in this tradition with the extension of the use of biologically inspired walking machines as agents. They are reasonably complex mechanical systems (many degrees of freedom) compared to wheeled robots, which have been used in most previous research. In addition, the creation of desired reactive behaviors has to be done using more advanced techniques.

However, there are many different techniques and approaches for robotic behavior control which can be classified into two main categories: one is deliberate control and the other is reactive control. According to R.C. Arkin (1998) [10], a robot employing deliberative reasoning requires relatively complete knowledge about the world and uses this knowledge to predict its actions, an ability that enables it to optimize its performance relative to its model of the world. This results in the possibility that the action may seriously err if the information that the reasoner uses is inaccurate or has changed since being first obtained. On the other hand, reactive control is a technique used for tightly coupling perception and action, and it requires no world model to perform the action of robots. In other words, this reactive system typically consists of a simple sensorimotor pair, where the sensory activity provides the information to satisfy the applicability of the motor response. Furthermore, it is suitable for generating robot behavior in the dynamic world. This means that robots can react to environmental stimuli as they perceive without concern for task planning algorithms or memory capacities.

In this book, we shall concentrate on the concept of reactive control to generate the behavior of four- and six-legged walking machines. In particular, we shall present a behavior controller based on a modular neural structure with an artificial neural network using discrete-time dynamics. It consists of two main modules: neural preprocessing and neural control[1] (Fig. 1.10).

The function of this kind of a neural controller is easier to analyze than many others which were developed for walking machines, for instance, by using evolutionary techniques [30, 72, 103, 119, 149, 168]. In general, they were too large to be mathematically analyzed in detail, in particular, if they used a massive recurrent connectivity structure. Furthermore, for most of these controllers, it is *hardly possible* to transfer them successfully onto walking machines of different types, or to generate different walking modes (e.g., forwards, backwards, turning left and right motions) *without modifying the network's internal parameters or structure* [22, 27, 56, 221].

In contrast, the controller developed here can be *successfully applied to a physical four-legged as well as to a six-legged walking machine,* and it is also able *to generate different walking modes without altering internal parameters or the structure of the controller.* Utilizing the modular neural structure, different reactive behavior controls can be created by coupling the neural control module with different neural preprocessing modules. Because the functionality of the modules is well understood, the reactive behavior controller of a less complex agent[2] (four-legged walking machine) can be applied also to a more complex agent (six-legged walking machine), and vice versa. A part of

---

[1] Here, *neural preprocessing* refers to the neural networks for sensory signal processing (or so-called neural signal processing). *Neural control* is defined as the neural networks that directly command motors of a robot (or so-called neural motor control). These definitions are used throughout this book.

[2] In this context, the complexity of an agent is determined by the number of degrees of freedom.

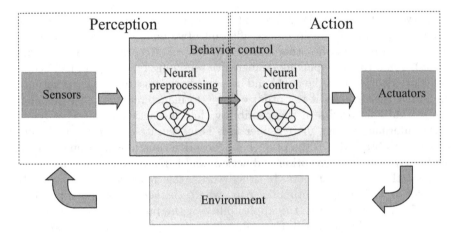

Fig. 1.10. The diagram of the modular reactive neural control (called behavior control). The controller acts as an artificial perception–action system, i.e., the sensor signals go through the neural preprocessing module into the neural control module which commands the actuators. As a result, the robot's behavior is generated by the interaction with its (dynamic) environment in the sensorimotor loop

the controller is developed by realizing dynamic properties of recurrent neural networks, and the other is generated and optimized through an evolutionary algorithm. On the one hand, the small recurrent neural networks (e.g., one or two neurons with recurrent connections [150, 151, 153]) exhibit several interesting dynamic properties which are capable of being applied to create the neural preprocessing and control for the approach used in this book. On the other hand, the applied evolutionary algorithm Evolution of Neural Systems by Stochastic Synthesis (ENS³) [97] tries to keep the network structure as small as possible with respect to the given fitness function. Additionally, every kind of connection in hidden and output layers, e.g., self-connections, excitatory and inhibitory connections, is also allowed during the evolutionary process. Consequently, the neural preprocessing and control can be formed using a small neural structure.

In order to physically build four- and six-legged walking machines for testing and demonstrating the capability of the behavior controllers, the morphologies of walking animals are used as inspiration for the design. The basic locomotion control of the walking machines is also created by determining the principle of animal locomotion. In addition, an animal's behavior as well as its sensing systems are also studied to obtain robot behavior together with its associated sensing systems. Inspired by the obstacle avoidance and escape behavior of scorpions and cockroaches, including their associated sensing systems, the behavior controller, called an "obstacle avoidance controller", and the sensing systems are built in a way that enables the walking machines to avoid obstacles or even escape from corners and deadlock situations. This behavior is represented as a negative tropism while a positive tropism is triggered by a sinusoidal sound at a low frequency—200 Hz. The sound induced behavior, in analogy to prey capture behavior of spiders, is called sound tropism. It is driven by a so-called sound tropism controller together with a corresponding sensory system. As a result, the walking machine reacts to a switched-on sound source (prey signal) by turning toward and finally making an approach (capturing a prey).

Eventually, all these different reactive behaviors are fused by using a sensor fusion technique[3] to obtain an effective behavior fusion controller, where different neural preprocessing modules have to cooperate. These reactive systems also aim to work as artificial perception–action systems in the sense that they perceive environmental stimuli (positive and negative tropism) and directly perform the corresponding actions. However, the created systems have no appropriate benchmarks for judging their success or failure. Thus, the ways to evaluate the systems are by empirical investigation and by actually observing their performance.

1.3 Organization of the Book

This chapter provided an overview of the research in the domain of agent–environment interactions, followed by the details of approaches to versatile artificial perception–action systems. The rest of this book is organized as follows:

Chapter 2 provides the biological background that served as an inspiration for the design of the reactive behaviors of walking machines, the physical sensing systems, the structures of walking machines and their locomotion control. It also shows how these biologically inspired systems are applied to the work done in this book.

Chapter 3 contains a short introduction to a biological neuron together with an artificial neuron model. Furthermore, it also describes, in detail, the discrete dynamical properties of a single neuron with a recurrent connection and an evolutionary algorithm. These are employed as the methods and tools used throughout this book.

Chapter 4 describes the biologically inspired sensory systems and walking machines which were originally built with physical components in this book. They serve as hardware platforms for experiments with the modular neural controllers or even as artificial perception–action systems.

---

[3] This fusion technique consists of two methods: a look-up table, which manages sensory input by referring to their predefined priorities, and a time scheduling method, which switches behavioral modes.

Chapter 5, which is the main contribution of this book, introduces the neural preprocessing of sensory signals and neural control for the locomotion of walking machines. It also presents different behavior controls which are the product of the combination between the different neural preprocessing units and the neural control unit. It ends up with the detail of behavior fusion control that combines all created reactive behaviors and leads to versatile artificial perception–action systems.

Chapter 6 shows the detailed results of the neural preprocessing tested with the simulated and real sensory signals. It also shows the capabilities of the controllers implemented on the physical walking machine(s) which generate different reactive behaviors.

Chapter 7 examines what has been achieved so far and suggests new avenues for further research.

2

Biologically Inspired Perception–Action Systems

Most of this book is devoted to creating and demonstrating so-called artificial perception–action systems inspired by biological sensing systems (perception) and animal behavior (action). Thus this chapter attempts to provide the biological background for understanding the approach taken in this book. It begins with a short introduction to some of the necessary principles of animal behavior. Then it concentrates on the obstacle avoidance and escape behavior of a scorpion and a cockroach, and continues with the prey capture behavior of a spider. Here, attention is given to the biological sensing systems used to trigger the described behaviors. Furthermore, different morphologies of walking animals are presented as inspiration for the design of walking machine platforms. Finally, a biologically inspired locomotion control, called a "central pattern generator" (CPG), is also discussed. This concept is later employed to generate the rhythmic leg movements of the machines.

2.1 Senses and Behavior of Animals

How can robotic behavior be designed, and how can it be created in a rational way? How can the desired behaviors cooperate? How are these behaviors applied to sensors and actuators? What kind of primitive behavior should be implemented in a robotic system, in particular in a mobile robot, before adding more complex behaviors? These are example questions which most roboticists always keep in mind before creating a robotic system that can interact with an environment. Therefore, all these questions must be answered somehow to provide the principal idea for creating the robot behavior as well as its physical system such as sensor and actuator types. A possible solution to these problems may be to observe and study animal behavior (actions) as well as sensing systems (perceptions), whereby they serve as inspiration for designs. It seems that animal behavior defines intelligence in the sense that an animal has the ability to improve its prospects of survival in the real world. From studying animal behaviors in their natural environment, ethologist roughly classified the behaviors into three major classes (from R.C. Arkin 1998 [10]).

- *Reflexes* are rapid stereotyped responses triggered by a certain environmental stimulus. The response perseveres as long as the stimulus is presented and depends upon the strength of the stimulus. Reflexes allow an animal to quickly adapt its behavior to unexpected environmental changes. Reflexes are usually employed for tasks such as postural control, withdrawal from painful stimuli and the adaptation of gait to uneven terrain.

- *Taxes* are orientational responses. These behaviors involve the orientation of an animal toward (positive tropism) or away (negative tropism) from a stimulus. Taxes occur in response to visual, chemical, mechanical and electromagnetic effects in a wide range of animals. For instance, a wandering spider exhibits positive tropism; that is, it orients to the airflow produced by a buzzing fly to capture the fly, which is known as "prey capture behavior" [18, 90]. Another kind of positive tropism is also evident in female crickets. They perform phonotaxis during courtship; that is, they turn into the direction of the calling of a male [137]. On the other hand, the negative tropism can be compared with, for example, an obstacle avoidance behavior during navigation or exploration in a scorpion [4, 202] as well as in an insect. They try to turn away from an obstacle which is perceived by their tactile sensing systems (e.g., hairs, antennas). However, the obstacle avoidance behavior can also be realized as part of the reflex response.

- *Fixed-action pattern* is a time-extended response pattern activated by a stimulus; i.e., the action perseveres for longer than the stimulus itself. The intensity and duration of the response are not controlled by the strength and duration of the stimulus. The triggering stimulus of a fixed-action pattern is usually more complex and specific than reflexes. In fact, once a fixed-action pattern has been activated, it will be performed even if the activating stimulus is removed. An example of a fixed-action pattern is the escape behavior of cockroaches. They immediately turn and run away when a predator attacks [172].

The easiest way to generate robotic behavior is perhaps by adopting animal behaviors described above. They are a reaction to an environmental stimulus perceived via the sensory system. Such a reaction is called a "reactive behavior". It can be used to express how a robot should react to its environment. To do so, a reactive robot system can also be clarified as a perception–action system; i.e., a robot perceives some environmental information and reacts to its environment without the use of background information or time history. This system is suitable for dynamic and hazardous environments because it responds directly to the environment that it senses.

Here, two distinctive reactive behaviors of animals were investigated, and associated sensing systems were focused upon. One is obstacle avoidance and escape behavior, which is represented by a negative tropism, while the other is a prey capture behavior, which acts as a positive tropism. Both behaviors are detailed in the following subsections; this information forms the basis for the design of the robot behavior and its physical sensing system in Chaps. 4 and 5.

2.1.1 Obstacle Avoidance Behavior

Obstacle avoidance behavior is realized in most animals because they are able to escape or avoid obstacles in cluttered real environments during the performance of an ordinary task (e.g., wandering around or seeking food). Indeed, if an animal is faced with an obstacle, it sometimes turns away from, climbs over, follows or even makes an exploration of the obstacle. These different behaviors usually depend on a situation and the property of the obstacle. The interesting parts of this desired behavior are how the animal senses the obstacle and which sensory system provides perceptual information of the obstacle. The biological evidence which supports these hypotheses is described below.

Scorpions are nocturnal and predatory animals that feed on a variety of insects, spiders, centipedes and even other scorpions. They have a poor visual system with difficulties in detecting obstacles or prey at long distances. Instead, it is mainly used as a photoreceptor for distinguishing between day or night [48]. Thus they mostly perceive environmental information via sensory hairs distributed over most parts of the body. For example, F.T. Abushama [4] observed that the scorpion *Leiurus quinquestriatus* uses the hairs on the distal-tarsal segments of the legs for humidity sensing while the pedipalps (the pincers), the pectines and the poison bulb appear to carry the hairs responsive to touch, odor and temperature, respectively (Fig. 2.1).

On the other hand, A. Twickel [202] observed the scorpion *Pandinus cavimanus* (Fig. 2.1) in the situation where the hairs on the pedipalps were used for collision detection. Here, the pedipalps play a role in the active perception of obstacles. Once the hairs on the first pedipalp collided with an obstacle, the scorpion started to slowly turn away from the side of the touch. During obstacle avoidance, it also performs a tactile exploration of the obstacle through the active pedipalps. Using the tactile hairs for obstacle perception, it is finally able to escape the obstacle. The series of photos of the obstacle avoidance behavior is presented in Fig. 2.2.

In analogy to obstacle avoidance behavior of scorpions, most insect species (e.g., crickets, cockroaches, stick insects, etc.) are also capable of escaping from an obstacle or even their predators. Some of them mainly perceive the information of obstacles or predators through antennal systems. The sensory system consists of two actively mobile antennas that project from the head of the insect, and are associated with the neural signal processing. Generally, insect antennas are exterior sensory structures composed of many tiny segments. They are highly sensitive to touch stimuli, and may even be able to discriminate textures [45]. They are flexible, and each of them can be swept

Fig. 2.1. The scorpion *Pandinus cavimanus* (modified from S.R. Petersen 2005 [160] and A. Twickel 2004 [202] with permission)

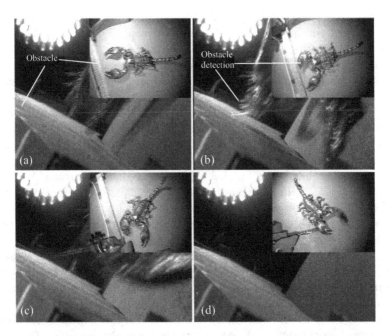

Fig. 2.2. Obstacle avoidance behavior of the scorpion *Pandinus cavimanus* (see from (**a**) to (**d**)). *Small windows* show the obstacle avoidance behavior while *large windows* present obstacle detection in close-up view. (Reproduced with permission of A. Twickel [202])

independently. The insect actively moves the antenna by controlling the muscle in specialized segments located at the base. An example of antennas in the female cricket *Gryllus bimaculatus* and the cockroach *Blaberus discoidalis* is shown in Fig. 2.3.

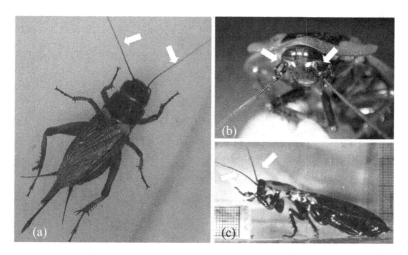

Fig. 2.3. The antennas of insects. (**a**) The female cricket where *arrows* indicate the position of antennas (reproduced with permission of T.P. Chapman [46]). (**b**) Front view of the cockroach where *arrows* indicate based segments for moving antennas. (**c**) Side view of the cockroach where *arrows* indicate the position of antennas (pictures (**b**) and (**c**) reproduced with permission of R.E. Ritzmann [163])

In fact, insect antennas appear to serve an amazing variety of tasks. For example, the use of antennas for chemical pheromone sensing has been suggested by D. Schneider in 1964 [183] and 1999 [184]. Antennas are also sensitive to air currents [31], which is found in the carrion beetle, while in cockroaches they have been determined for wind-mediated escape [25, 194]. Particularly, through the sense of touch to be required for acting as mechanoreceptor, they can probe for foothold in rough terrain [79] and actively explore it during walking [62] (e.g., antennas of stick insects). Furthermore, they are used for wall-following [45] and even for touch-evoked behavior[1] [50, 51, 181] (e.g., antennas of cockroaches).

Ideally, the work here would concentrate on touch-evoked behavior in cockroaches to understand how they react to touch stimuli through the antennas. Touch-evoked behavior was precisely investigated by C.M. Comer et al. [49, 51]. There, the reaction of a cockroach with a predatory wolf spider was

---

[1] The context implies to "an obstacle avoidance behavior" if antennas make contact with an obstacle or to "a predator escape behavior" if stimulus is produced by an attack of its predator, e.g., a wolf spider.

attempted. In the observed situation, at the beginning before the antennas made contact with the spider, the spider was in motion toward the cockroach while the cockroach was standing still. After that the right antenna touched the striking spider and the cockroach started to turn to the left. Finally, the cockroach was able to escape from its predator. The described behavior [49] is shown in Fig. 2.4.

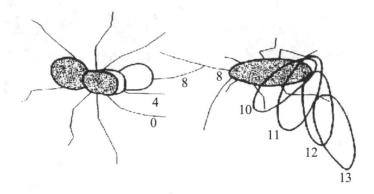

Fig. 2.4. A predator escape behavior of a cockroach, where the *shaded outlines* indicate animal's initial position (the spider on the left and the cockroach on the right); *numbers* present position at successive video frames. They came into contact on frame 8, and the cockroach began to turn and was able to escape at the end (see from frame 8 to 13). The figure is taken from C.M. Comer et al. [49]

Other experiments to observe the evasive behavior (escape behavior) of cockroaches are reported in [49]. The behavior was triggered by artificial touch stimuli at one antenna. The resulting response was that cockroaches mostly oriented away from the side of touch with an average vector suitable for escape. In this situation, the cockroaches turned to the left side when the right antenna was tapped. The orientation of turns is summarized as a circular histogram shown in Fig. 2.5, and extremely short latency was observed where the mean latency of the turn was 33 ms.

From the investigation above, the obstacle avoidance and escape behavior of a scorpion and a cockroach can be determined as reactions to a negative tropism. Such reactions are also standard. Animals actually turn away from the side on which contact is made using their sensing systems. This negative tropism will be later taken into account for the behavior control of walking machines in the way that the machines will turn away from the side of the stimulus (e.g., obstacle detection). Another aspect from this biological investigation states that biological sensory systems (e.g., the tactile hairs of the scorpion and the antennas of the insects) at the anatomical level are somewhat complicated. Thus, no attempt to model the detailed anatomy of these

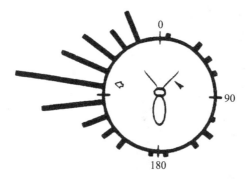

Fig. 2.5. The circular histogram shows the orientation of turns which respond to touch stimuli. The *open arrow* indicates the average angle of turning while the black one represents the position where the right antenna has been touched. The figure is taken from C.M. Comer et al. [49]

sensors is done in this book. Instead, physical sensors associated with their neural preprocessing will be modeled in a simple way with respect to the functionality of biological sensory systems.

2.1.2 Prey Capture Behavior

All spiders are really hairy creatures, and most spiders have very poor eyesight. Thus, they mostly rely on their hairs for sensing their environment instead of their eyes. The hair is used to perform a surprising variety of tasks (Fig. 2.6, right). For example, tactile-sensitive hairs on the legs help the spider to move freely around its terrain [186, 187]. There are also airflow-sensitive hairs which are important for detecting its prey [16, 17]. Furthermore, the hairs on the pedipalps are used as chemoreceptors which are sensitive to taste and odor [60, 61] and also associated with mate recognition [76, 147].

From concise investigation on sensing systems of spiders, several attractive functions of the hairs have been mentioned. By now, the airflow-sensitive hairs called *Trichobothria* of the spider *Cupiennius salei* (Fig. 2.6, left) are well researched. Actually, trichobothria are the sensillum (sense organ), having the hair-like structure which arises from a socket in the cuticle. They have low mass and are very flexible. Thus, they are extremely sensitive to the airflow stimulus and the auditory cues[2] in a low-frequency range between

---

[2] In 1883, F. Dahl found that the trichobothria respond to low tones produced by bowing a violin, and thus he classified them as "auditory hairs" [57]. Later on, in 1917, H.J. Hansen published an article describing sensory organs in arachnida where auditory hairs were also mentioned [87].

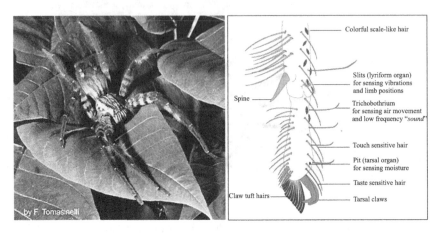

Labels in figure:
Colorful scale-like hair
Slits (lyriform organ) for sensing vibrations and limb positions
Spine
Trichobothrium for sensing air movement and low frequency "sound"
Touch sensitive hair
Pit (tarsal organ) for sensing moisture
Taste sensitive hair
Claw tuft hairs
Tarsal claws
by F. Tomasinelli

Fig. 2.6. *Left:* the wandering spider *Cupiennius salei* (Copyright 2002 by F. Tomasinelli and reproduced with permission [200]). *Right:* the layout of the hairs on the spider leg (Copyright 2002 by Australian Museum and reproduced with permission [1])

approximately 40 and 600 Hz [16, 19]. Through the use of these hairs, the spider is able to detect its prey (e.g., a buzzing fly) which generates the airflow at a frequency range of around 100 Hz. In other words, this sensing system (the hairs together with associated neural signal processing) acts as a matching filter. It reacts to the biologically significant signals (e.g., prey signals) while it filters out surrounding noise as well as interfering signals (e.g., background airflow). This is because most background noise has a very low frequency (a few hertz), which nicely contrasts with the frequency of prey signals [15]. Figure 2.7 shows the response of the individual hair to the prey signal (airflow generated by a buzzing fly during stationary flight 5 cm away).

In fact, the spider *Cupiennius salei* has approximately 950 tirchobothria with the length up to 1400 μm which are located on the tarsus, the metatarsus and the tibia of the spider leg (Fig. 2.7). This sensing system is adequate to perform "prey capture behavior" when it is stimulated by the airflow generated by a buzzing fly at a distance of up to approximately 30 cm. As a result, the spider orients its movement toward the direction of the stimulus and then jumps to the targeted buzzing fly [18, 35, 90]. The series of photos of prey capture behavior is shown in Fig. 2.8.

As shown here, the prey capture behavior represents a positive tropism. Such reaction is mostly found in predatory animals, e.g., spiders, scorpions and so on. They respond to a prey stimulus through sensing systems, e.g., sensory hairs. Consequently, they turn in the direction of the stimulus source and then try to capture a targeted prey. These kinds of a positive tropism and the described sensing system (trichobothria) of the spider are able to be reproduced on a walking machine (in an abstract form) whereby a puff of

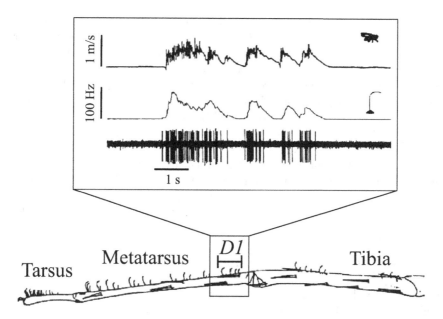

Fig. 2.7. The recorded signal of the trichobothria at location D1 on the leg of the spider *Cupiennius salei* in response to the airflow produced by a stationary buzzing fly (modified from F.G. Barth 2002, p. 253 [15])

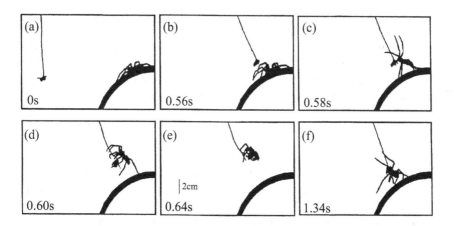

Fig. 2.8. The spider *Cupiennius salei* jumps toward a buzzing fly on a leash (see from (**a**) to (**f**)). The action time is indicated on the *lower left corner* of each photo (modified from F.G. Barth 2002, p. 257 [15])

the wind, which is normally generated by the buzzing fly, is replaced by a low-frequency sound around 200 Hz. Also, the biological airflow detectors are simplified to physical sound detectors instead. Thus, as a result, the physical sound detectors together with associated artificial neural preprocessing[3] shall enable the walking machine to react to a switched-on sound source by turning toward and making an approach to it at the end (like capturing prey). This sound-induced behavior is called "sound tropism".

Eventually, these different reactive behaviors will be integrated into a behavior controller of a walking machine, where the controller has to cooperate as in a versatile perception–action system. For example, the stimulus through antenna-like sensors generates a negative tropism while the low-frequency sound triggers a positive tropism, so that the walking machine (i.e., a predator) follows a switch-on sound source (i.e., a prey signal) but avoids obstacles.

2.2 Morphologies of Walking Animals

In order to explore the neural control of the biologically inspired behaviors in a physical agent, the specific agent's body must be carefully designed because it defines the possible interactions with its environment. In addition, the body of the agent also determines the boundary conditions of an environment in which it can operate successfully. The design of the neural control depends on the morphology of the agent, i.e., the type and position of the sensors and the configurations of the actuators. Choosing too simple a design, the behavior of the body may be of limited interest and it may obstruct the need for an effective neural control for a complex system. To achieve this potential, agents having morphologies similar to walking animals are preferred. In other words, biologically inspired walking machines are the robot platforms for the approach of this work. Such machines are more attractive because they can behave somewhat like animals and they are still a challenge for locomotion control.

Two walking animals were observed to benefit the leg and trunk designs of four- and six-legged walking machines (physical agents). The inspiration for the structure of a four-legged walking machine came from the biological principles that a salamander uses to obtain an efficient walking pattern [34, 165], while the design of the legs and the trunk of a six-legged walking machine follows the way that a cockroach walks and climbs [163, 216]. The details of the morphologies of both walking animals are described below.

---

[3] This physical sensing system together with its neural preprocessing shall perform like trichobothria with associated biological neural processing (a matched filter). That is, the physical sensors detect the signal while the neural preprocessing acts as a matched filter passing only the low-frequency sound (200 Hz) to trigger a so-called sound tropism.

2.2.1 A Salamander

A salamander is a vertebrate walking animal that belongs to the group of amphibian tetrapod. It is able to traverse both on land and water. It has small limbs projecting from its trunk for walking on land. Each limb is formed of three main segments which are thigh, shank and foot (Fig. 2.9).

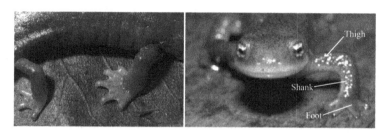

Fig. 2.9. The limbs of a salamander. (Reproduced with permission of G. Nafis 2005 [139] (*left*) and K. Grayson 2000 [83] (*right*))

All limbs are quite small and far from each other, causing difficulty for locomotion, in particular on land. Therefore, it also uses the movement of the trunk bending back and forth coordinated with the movements of the limbs for an efficient walking pattern [100]. The trunk is mainly created from muscles propagating along the backbone (musculature). This musculature has the advantage of more flexible and faster motions and aids in climbing. Generally, during locomotion on land, its trunk bends to one side causing an increase in the step length of the two diagonally opposite lifted limbs which are pushed forward while the other two limbs are pushed backward simultaneously. As a result, it performs a trot gait. The locomotion of a salamander on land is presented in the series of photos in Fig. 2.10.

Fig. 2.10. The locomotion of a salamander (from *left* to *right*). An *open circle* of each photo assumes to a backbone joint which connects the first segment (*1*) with the second segment (*2*) and makes an active bending movement of a trunk for locomotion. (Courtesy of J.S. Kauer [108] (Kauer Lab at Tufts University))

By realizing the salamander structure, the trunk of a four-legged walking machine was designed with a backbone joint which can rotate in a vertical axis. The joint then facilitates more flexible and faster motions like the movement of a salamander. Each leg is modeled more simply than a salamander leg but still maintains the operations of a salamander leg; i.e., it can perform forward–backward and up–down motions (see in Sect. 4.2).

2.2.2 A Cockroach

A cockroach is an invertebrate walking animal in the phylum of arthropods. It has six legs, and each leg is composed of multiple segments: coxa, trochanter, femur, tibia and tarsus (foot). The upper leg segments generally point upwards and the lower segments downwards. The legs project out from the trunk like a salamander. They are oriented around its trunk in a way that the two front legs point forwards while the four rear legs typically point backwards to maintain stability in walking (Fig. 2.11). Such orientation can be beneficial in climbing over an obstacle; i.e., a cockroach can easily move its front legs forward to reach the top of an obstacle while the rear legs power its motion by rising its trunk up and pushing it forward. As a result, it can climb over the obstacle (Fig. 2.3c) [216]. Moreover, front legs are also used to detect stimulus coming from the front while rear legs perceive stimulus from the back.

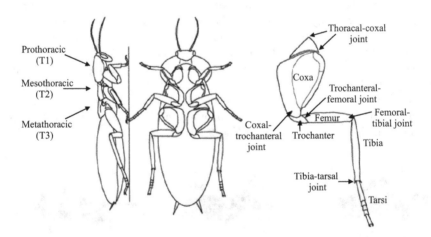

Fig. 2.11. The legs of a cockroach and the orientation of legs around its trunk (modified from J.T. Watson et al. 2002 [211])

Concerning its number of legs, a cockroach normally performs a typical tripod gait for the walking pattern, where the front and rear legs on one side together with the middle leg on the other side support the trunk (so-called stance phase) simultaneously while the other three legs are in the air (so-called swing phase). These support legs are then replaced by the other three legs in the next step. While walking under normal conditions, its trunk does not bend back and forth like a salamander as the structure of its trunk is different from a salamander. The trunk is not formed by a muscle. Instead, it consists of three main segments: prothoracic (T1), mesothoracic (T2) and metathoracic (T3) (Fig. 2.11, left). This structure is advantageous for climbing on an obstacle, as it has the transition between vertical and horizontal surface. It can bend its trunk downward at the joint between the first (T1) and second (T2) thoracic segments to keep the legs close to the top surface of the obstacle for an optimum climbing position and even to prevent unstable actions (Fig. 2.12).

Fig. 2.12. A cockroach climbing over a large obstacle block (adapted from R.E. Ritzmann 2004 [163])

Inspired by the morphology of a cockroach, the trunk of a six-legged walking machine was constructed with a backbone joint rotating in a horizontal axis. Thus, the backbone joint is like the connection between the first and second thoracic segments of a cockroach. It will provide enough movement for the machine to climb over an obstacle by rearing the front legs up to reach the top of an obstacle and then bending them downward during step climbing. Each leg was designed with respect to the movement of a cockroach leg. It consists of three joints, where the first joint can move the leg forward and backward, the second one can move the leg up and down, and the last one is for elevation and depression or even for extension and flexion of the leg (see more details in Sect. 4.2).

2.3 Locomotion Control of Walking Animals

The basic locomotion and rhythm of stepping in walking animals mostly relies on a CPG [55, 84, 158, 193]. The CPG is a group of interconnected neurons that can be activated to generate a motor pattern without the requirement of sensory feedback [58, 94]. The evidence which supports this hypothesis was originally demonstrated by T.G. Brown in 1911 [42]. He discovered that the rhythmic patterned activity of leg muscles in a cat, similar to those that appear during walking, could be activated although all input from sensory nerves in the legs had been eliminated. This is because the processes underlying cat locomotion are situated in the spinal cord; i.e., if the dorsal roots[4] of a cat are cut, the ventral roots[5] are still able to induce a rhythmic patterned activity (Fig. 2.13). Later on, in 1966, M.L. Shik et al. [188] presented that cats without the higher levels of the nervous system (the cerebral hemispheres and the upper brain stem (Fig. 2.13)) are still able to walk in a controlled manner on a treadmill. This result has been accumulated to support the original proposal of T.G. Brown that the basic rhythmic movements in each leg of the cat can be generated without sensory input.

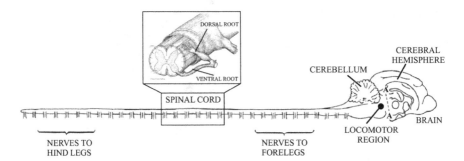

Fig. 2.13. The spinal cord and lower brain stem of a cat are cut from the cerebral hemispheres and the upper brain stem at the cross section A' - A (modified from K.G. Pearson 1976 [158] and Copyright 2005 by Pearson Education with permission)

Moreover, S. Grillner and his colleagues [85, 148] made experiments on the pattern of activity in the flexor and extensor muscles of cats after the elimination of sensory input from the receptors in the legs. They found that the rhythmic patterned activities in flexors and extensors of the cat's hind leg could still be generated although the spinal cord was cut from the hind-leg segments. This important result leads to the discovery that the rhythmic

---

[4] The two nerve fiber bundles of a spinal nerve that carries sensory information to the central nervous system.

[5] The part of a spinal nerve, consisting of motor fibers, that arises from the previous section of the spinal cord.

patterned activity is generated not only by the spinal cord but also with the effect from a central rhythm generator for each leg. Similar results were obtained by K.G. Pearson and J.F. Iles [158, 159] in studies of the cockroach. After disconnecting all sensory input from the legs, the rhythmic patterned activities in hind-leg flexor and extensor motor neurons remain functional. The examples of the rhythmic patterned activities after all sensory input from receptors in the hind leg of the cat and the cockroach had been eliminated are shown in Fig. 2.14.

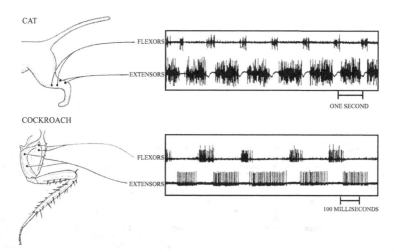

Fig. 2.14. The existence of a central rhythm generator for each leg of the cat and the cockroach is corroborated by the fact that even after all sensory input from receptors in the hind legs had been eliminated, they could still work. The rhythmic bursts of electrical activity were generated in the flexor and extensor muscles in the hind legs of both animals (adapted from K.G. Pearson 1976 [158])

The rhythmic patterned activities of the CPGs together with the mechanism that coordinates the motion of all legs form the basic walking patterns [44]. In the cat, there are four basic patterns (walking gaits): the walk, the trot, the pace and the gallop. During walking, trotting and pacing the movements of the two hind legs are out of phase as are the movements of the two forelegs. The difference between the three gaits is the timing of the stepping of the two legs on each side of the animal. For example, during slow walking, the left foreleg steps shortly after the left hind leg and before the right hind leg. The stepping sequence is the following: left hind leg, left foreleg, right hind leg, and right foreleg, and so on. When the walking speed is increased until the diagonal legs step at the same time, then the animal is trotting.

Pacing is realized by the simultaneous stepping of the two legs on one side. As a result, the animal can move with slightly higher speed than trotting. The fastest movement of the animal is galloping where the opposite legs move almost synchronously and the forelegs are out of phase with the hind legs. In the cockroach, which of course has six legs, the walking patterns can instead be simply determined by the walking speed. For fast walking gait, the animal is always supported by at least three legs; e.g., the left rear, right middle and left front legs step in phase while the remaining legs step out of phase. For that reason, the gait is called the tripod gait. If the walking speed decreases, the gait is changed and it can be described as a sequence of the three legs on each side moving from the back to the front. The basic walking gaits of the cat and the cockroach are presented in Fig. 2.15.

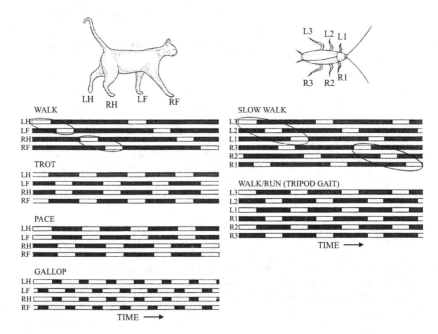

Fig. 2.15. Walking gaits of the cat and the cockroach are depicted from *left* to *right*. Each *white block* indicates that the foot has no ground contact (swing phase), while each *black block* indicates that the foot touches the ground (stance phase). During slow walking, there is a back-to-front sequence of stepping for both animals; the sequence are marked by the *ellipses* (modified from K.G. Pearson 1976 [158])

The CPGs mentioned above seem to underly the production of all basic rhythmic walking gaits. This does not mean that sensory inputs are unimportant in the patterning of locomotion. In fact, the sensory input also plays an important role, to change the walking patterns and animal behaviors. For example, animals use the sensory feedback from the moving legs to adapt the walking patterns for irregular terrain [21, 79]. It is even used to produce well-coordinated motor patterns which are mostly found in stick insects [20, 44]. In addition, the sensory input also controls animal behavior responding to environmental stimuli, and it, of course, effects to generate the appropriate motions. Thus, on the one hand, the basic locomotion or the rhythmic patterned activity of legs is generated by the CPGs performing as a low level control while, on the other hand, the sensory input acting as a high level control will command for different walking patterns, e.g., changing a walking gait from slow to fast walking or vice versus as well as changing the walking directions.

From the biologically inspired locomotion control, the basic rhythmic movements of the legs of the four- and six-legged walking machines will be basically generated by the CPG and the sensory information will be also used to modify the leg movements to obtain the various walking patterns. Consequently, the walking machines shall normally walk with the trot gait for four legs and the tripod gait for six legs, and the sensory inputs will steer the walking directions of the machines in turning left, right and even walking backward (see more details in Sect. 5.2).

2.4 Conclusion

Animals are excellent models for the design of robotic systems. They show fascinating behaviors which can serve as an inspiration for modeling the behavior control of the walking machines including their physical sensing systems. Generally, animals respond directly to their environment through their senses. This reaction is defined as a reactive behavior and it is the basis to express how the walking machines should react to their environment. In this book, different reactive behaviors together with the associated sensing systems are investigated, whereby one is an obstacle avoidance behavior represented as the negative tropism and the other is a prey capture behavior classified as positive tropism. Both behaviors are to be emulated in abstract forms in our walking machine(s).

We also tried to simulate the morphologies of walking animals. The morphologies of a salamander and a cockroach were taken into account so as to benefit the leg and trunk designs of the four- and six-legged walking machines, especially their use of the backbone joint or the interconnection joint between segments for efficient locomotion. Furthermore, the basic locomotion control of the walking animals was also studied. It mostly relies on a CPG, which is the group of interconnected neurons producing rhythmic patterned outputs without the requirement of sensory feedback. Thus, the locomotion control of the two walking machines will be basically generated by realizing the concept of the CPG, and it will then be modified by sensory signals with respect to environmental stimuli.

3

Neural Concepts and Modeling

This chapter presents methods and tools which are to be used throughout this book. It starts with a short introduction to a biological neuron together with an artificial neuron which is followed by the comparison of network structures between feedforward and recurrent neural networks. Then the discrete-time dynamical properties of the single neuron with a recurrent connection are described. Finally, artificial evolution is presented as a tool to develop and optimize neural structures as well as the strength of synapses.

3.1 Neural Networks

As illustrated by research which applied artificial neural networks (ANNs) to a wide field of applications, e.g., signal processing [123, 127, 196, 205, 225], robot control [23, 29, 56, 100, 114, 128, 156, 219], robot learning [14, 73, 82, 182, 220], etc., neural networks have the capability to deal with many kinds of problems including nonlinear problems. Moreover, there are various reasons for using neural networks for the work in this book. First, they are based on biological neural processing systems. Therefore, they are parallel-distributed processing patterns; i.e., their structure can consist of a very large number of synapses and neurons that can convey and process information simultaneously with a strong fault-tolerant behavior. In other words, many synapses or neurons must be damaged before the overall neural network system stops working properly. Second, they have a number of excellent properties; i.e., they are robust, they can be adaptive if a suitable on-line learning method is designed, they have the ability to handle small variations of noise and they even exhibit dynamical behavior (oscillatory, hysteresis, chaotic patterns, etc.), in particular recurrent neural networks (RNNs). And last, which is of relevance in this book, is that they are able to build a robot brain as a composition of different neural modules interacting in a cooperative or competitive way to produce the desired robot behavior. This means that ANNs can extend an existing neural system to improve the robot's behavior or even to obtain a robust behavior.

3.1.1 A Biological Neuron

In this section, a biological neuron is briefly discussed in order to provide a basic idea of its structure and principal functions. Thus, the physiological processes are not detailed here but are discussed further in the following references: J.A. Anderson (1995) [7], R. Hecht-Nielsen (1990) [142] and R. Rojas (1996) [174].

The human brain is the most complex structure known in the universe. It consists of approximately 10^{11} neurons, which are highly interconnected, and they communicate through a connection network having a density of approximately 10^4 synapses per neuron. This produces approximately 10^{14} synapses in the whole network. Figure 3.1 shows a model of the biological neuron consisting of four main components: the dendrites, the cell body called "soma", the axon and the synapses.

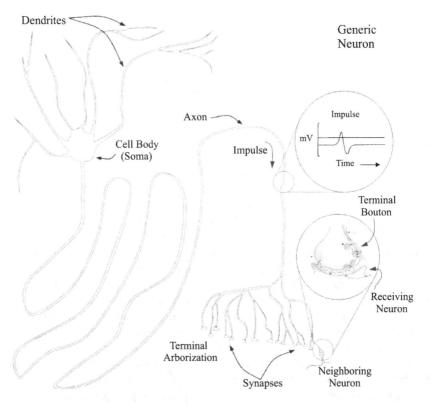

Fig. 3.1. A diagram of the generic neuron and a sample of an electrical impulse (modified from J.A. Anderson 1995, p. 7 [7])

Dendrites transmit information from other neurons to the soma. The axon makes connections to other neurons via synapses. Synapses can be excitatory if they cause firing in the form of spikes (increasing the activation level of a neuron) or they can be inhibitory if they prevent the firing of the response (decreasing the activation level of a neuron). The firing condition occurs when the excitation exceeds the inhibition by the amount called the threshold of the neuron, typically a value of roughly $+40\,\text{mV}$ [9]. Since a synaptic connection causes the excitatory or inhibitory reactions, it is useful to assign positive and negative weight values, respectively, to such connections.

However, there are a large number of various types of real neurons in the human brain, and they have also different dendritic shapes. Examples of real neurons are shown in Fig. 3.2.

3.1.2 An Artificial Neuron

A biological neuron has a high complexity in its structure and function; thus, it can be modeled at various levels of detail. If one tried to simulate an artificial neuron model similar to the biological neuron, it would be impossible to work with. Hence an artificial neuron has to be created in an abstract form which still provides the main features of the biological neuron. In the abstract form for this approach, it is simulated in discrete time steps and a neural spiking frequency (or called a firing rate)[1] is reduced to only the average firing rate. It is given by one simple output value. Moreover, the amount of time that a signal travels along the axon is neglected.

Before describing the artificial neural model in more detail, one can compare the correspondence between the respective properties of biological neurons in the nervous system and abstract neural networks to see how the biological neuron is transformed into the abstract one. This comparison is shown in Table 3.1 (from R. Pfeifer and C. Scheier 1999 [161]).

Table 3.1. Comparison of biological and artificial neurons

Nervous system	Artificial neural network
Neuron	Processing element, node, artificial neuron, abstract neuron
Dendrites	Incoming connections
Cell body (Soma)	Activation level, activation function, transfer function, output function
Spike	Output of a node
Axon	Connection to other neurons
Synapses	Connection strengths or multiplicative weights
Spike propagation	Propagation rule

---

[1] The number of spikes that a neuron produces per second.

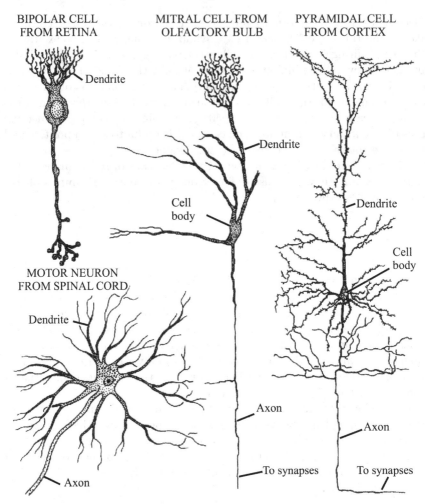

BIPOLAR CELL
FROM RETINA

MITRAL CELL FROM
OLFACTORY BULB

PYRAMIDAL CELL
FROM CORTEX

Dendrite

Dendrite

Dendrite

Cell
body

Cell
body

MOTOR NEURON
FROM SPINAL CORD

Dendrite

Axon

Axon

To synapses

Axon

To synapses

Fig. 3.2. Four different types of biological neurons are shown, each specialized for the specific function which they perform (from S.W. Kuffler et al. 1984, p. 10 [116])

The structure of a standard additive neuron model is shown in Fig. 3.3. This neural structure together with the given activation function and transfer function is employed throughout this book.

All weighted inputs (coming from sensors or other neurons, indicated by o_j) and a bias term used as a fixed input b_i are simply summed up and passed through an activation function to produce a level of activation. Therefore, the activation function of the standard additive neuron is given by:

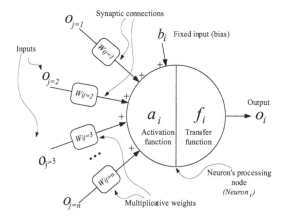

Fig. 3.3. The structure of an artificial neuron. Each neuron can have multiple input connections, which can originate from other neurons or from a sensor, but there is only one output signal. Then the single output signal can be distributed in parallel (in other words, multiple connections carrying the same signal) to other neurons or to an external system, e.g., a motor system

$$a_i = \sum_{j=1}^{n} w_{ij} o_j + b_i, \quad i = 1, \ldots, n, \tag{3.1}$$

where a_i is the activity of neuron i, n denotes the number of units, w_{ij} represents the synaptic strength or weight of the connection from neuron j to neuron i, b_i refers to a fixed internal bias term together with a stationary input to neuron i, and o_j is the input(s). In discrete time steps, the activation is then updated at each time step t, defined as an integer value. Thus, the activation function given in Eq. (3.1) can be rewritten as:

$$a_i(t+1) = \sum_{j=1}^{n} w_{ij} o_j(t) + b_i, \quad i = 1, \ldots, n. \tag{3.2}$$

The activation function is then transformed by a transfer function f_i to obtain a neuron output o_i. The most widely used transfer functions are shown in Fig. 3.4.

The linear threshold transfer function is similar to a step function. It sums the inputs and the activation level of the neuron is inactive (zero or -1) until the threshold value Θ is reached, at which point the neuron becomes active ($+1$) (Fig. 3.4a). For the linear transfer function (Fig. 3.4b), it simply sums the input, and it is often used as a buffer between external input signals (e.g., coming from sensors) and the determined network. The last transfer function which has been commonly used in a neural network model is the sigmoid or logistic transfer function (Fig. 3.4c). It is a smoothed version of a step function. Its output value is around zero (or ≈ -1), at a lower bound for low

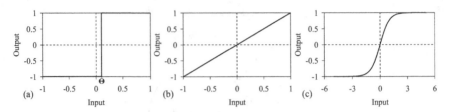

Fig. 3.4. (a)–(c) Most widely used transfer functions. (a) Linear threshold transfer function. (b) Linear transfer function. (c) Nonlinear sigmoid transfer function

input. At some point, it begins to increase rapidly before saturating ($\approx +1$ for an upper bound) at higher levels of input. The sigmoid transfer function with a lower bound at ≈ -1, called the "hyperbolic transfer function" ($\tanh(x)$), is used throughout the rest of the book because it is more convenient to apply in controlling a robot. It can also be justified by the observation that many biological neurons have a nonzero spontaneous firing rate. The equation of this transfer function is given by:

$$f(a_i) = \tanh(a_i) = \frac{2}{1 + e^{-2a_i}} - 1. \tag{3.3}$$

This transfer function is bounded between -1 and $+1$. Its boundary can be interpreted as the summation of synaptic inputs at the dendrites and cell body level in biological neurons. By applying the sigmoid transfer function, the neuron output o_i is determined as follows:

$$f(a_i) = \tanh\left(\sum_{j=1}^{n} w_{ij} o_j + b_i\right). \tag{3.4}$$

In any case, if the inputs to neuron i come from other neurons (i.e., outputs of neuron j) instead of sensors, the activation function of the standard additive neuron in the discrete-time domain can be described by:

$$a_i(t+1) = \sum_{j=1}^{n} w_{ij} \tanh(a_j(t)) + b_i, \quad i = 1, \ldots, n. \tag{3.5}$$

3.1.3 Models of Artificial Neural Networks

The arrangement of artificial neurons and their interconnections can have a profound effect on the processing capabilities of the neural networks. In general, all neural networks have a set of neurons receiving inputs from the outside world (e.g., sensor data). This set is indicated as the "input neurons". Many neural networks also have one or more internal neurons called the "hidden neurons" which receive inputs from other neurons or themselves. The set of neurons that represent the final result of the neural network, which is sent out to control external devices (e.g., motors), is defined as the "output neurons". Sets of neurons that have similar characteristics and are connected to other neurons in similar ways are called "layers" or "slabs".

Concerning connection topologies that define the direction of data flows between the input, hidden and output neurons, these can be classified into two different types of network architectures, so-called feedforward network and recurrent network. A feedforward network has a layered structure. Each layer consists of neurons which receive their input from neurons in a layer directly below and send their output to neurons in a layer directly above. This network does not have internal feedback; in other words, there exist only forward connections which produce forward activities of neurons. Therefore, feedforward networks are static; that is, their output depends only on the current inputs and the networks represent just simple nonlinear input–output mappings (Fig. 3.5, left).

On the contrary, if feedback exists within the connection structure which allows cyclic propagation of activity (or backward activities), the network is called a recurrent network (Fig. 3.5, right). The output of the network depends on the past inputs; thus, the network can represent various dynamical properties (e.g., hysteresis, oscillation and even deterministic chaos). Some dynamical behaviors of the network are useful for signal processing and robot control being the approach of the book. Therefore, this book is concentrated on applying recurrent neural networks together with their dynamical behavior to create so-called versatile artificial perception–action systems described in Chap. 5. However, there are two exceptions of applying the network to create the neural controller of the system, where input neurons can receive only input from the outside world (sensor data) and the number of the input and output neurons is determined by the number of sensors and motors used, respectively.

3.2 Discrete Dynamics of the Single Neuron

The single neuron with a self-connection, namely a recurrent neuro-module, has several interesting (discrete) dynamical properties which have been investigated by F. Pasemann [151, 152] and others [13, 88, 89, 170]. From these investigations, the single neuron with an excitatory self-connection has a hysteresis effect while the stable oscillation with period-doubling orbit can be observed for an inhibitory self-connection. However, both phenomena occur for specific parameter domains of an input and a self-connection weight.

In this book, the hysteresis effect is utilized for preprocessing sensor signals as well as robot control (described in Sect. 5.1). By now, the recapitulation of the used dynamical property of a recurrent neuro-module is discussed by employing the single neuron model presented in the previous section. The corresponding dynamics is parameterized by the input I and the self-connection w. The discrete dynamics of the single neuron with a self-connection is given by:

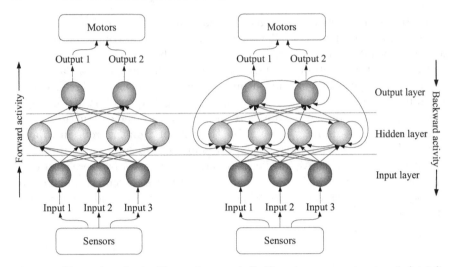

Fig. 3.5. Examples of a feedforward network (*left*) and a recurrent network (*right*). Generally for robot control, the input for the networks comes from the sensors while their output is sent to control the motors

$$a(t+1) = wf(a(t)) + \theta, \tag{3.6}$$

with the hyperbolic transfer function

$$f(a) = \tanh(a) = \frac{2}{1 + e^{-2a}} - 1, \tag{3.7}$$

where the parameter θ stands for the sum of the fixed bias term b and the variable total input I of the neuron. The model neuron with a self-connection for the investigation is presented in Fig. 3.6.

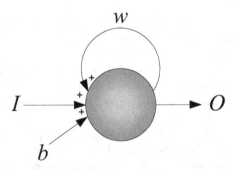

Fig. 3.6. The model neuron with a self-connection

As mention above, the hysteresis effect is observed only for an excitatory self-connection with the specific parameter domains; thus, the dynamics of an excitatory self-connection is shown here while the dynamics of an inhibitory self-connection as well as the detail of mathematical proof are referred to [151, 152].

By simulating the dynamical behavior of varying the excitatory self-connection w together with the input θ, two different domains in the (θ, w)-space are observed (Fig. 3.7).

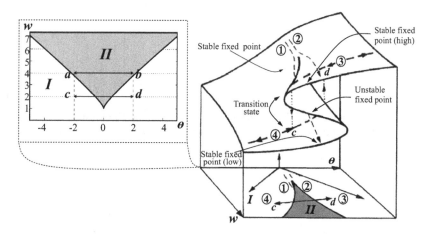

Fig. 3.7. The dynamics of a neuron with an excitatory self-connection. *Left*: Parameter domains for one stable fixed point (I), two stable and one unstable fixed points (II). *Right*: The cusp catastrophe with respect to the dynamics of the neuron. One can compare between the *left* and *right diagrams* to obviously see two stable fixed points and one unstable fixed point which exist in region II. There are also transition states shown on the *right diagram* where the system changes from one (low) stable fixed point to another (high) stable fixed point and vice versa (F. Pasemann 2005, personal communication)

In region I, there exists a unique stable equilibrium (one fixed point attractor) for the system while three stationary states (one unstable fixed point and two coexisting fixed point attractors which are the low and high points) are found in region II. In fact, the hysteresis effect of the output appears when the input θ crosses the region I and II; e.g., θ sweeps over the input interval between -2 and $+2$ for a fixed $w = 2$ (see also an *arrow line* between c and d in Fig. 3.7). If w is increased to 4 while θ still varies over the input interval (between -2 and $+2$); i.e., θ does not pass back and forth through the region I and II (see also an *arrow line* between a and b in Fig. 3.7). Instead, it varies inside the region II; consequently, the output O will stay at one fixed point attractor (either high or low fixed point attractor) depending on where the input starts. Furthermore, the width of the hysteresis loop is defined by the strength of the self-connection $w > +1$; i.e., the stronger the self-connection, the wider the loop is [96]. The comparison of the width of the hysteresis loop is presented in Fig. 3.8.

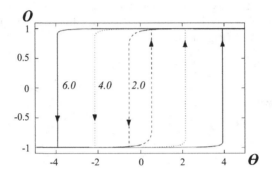

Fig. 3.8. Comparison of the "hysteresis effects" between the input θ and output O for $w = 2.0$, 4.0 and 6.0, respectively (F. Pasemann 2005, personal communication)

One can utilize such different sizes of a hysteresis loop for robot control, e.g., the turning angle of a mobile robot for avoiding obstacles can be determined by the width of a hysteresis loop. In other words, the wider the loop, the larger the turning angle is (see also Sect. 5.1.3). Additionally, there is an example situation shown in Fig. 3.9 where the hysteresis effect depends on the frequency of a dynamic input, e.g., slowly and quickly varying inputs.

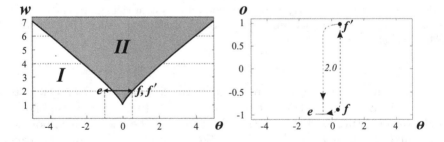

Fig. 3.9. The example of the dynamics of the recurrent neuron with the dynamic input. *Left*: The dynamic input θ varies between ≈ -1 which is in region I and ≈ 0.5 which is near to the border where the system can jump from one fixed point (low) to another fixed point (high). *Right*: The hysteresis effect of the dynamic input for a fixed $w = 2$, see text for details (F. Pasemann 2005, personal communication)

In this situation, the hysteresis will appear if the dynamic input has low frequencies; i.e., the system will move from point e to f' through f in one path and it will return to point e again in another path, resulting in the hysteresis loop. On the other hand, if the frequency of the dynamic input is high, the system will move from point e to point f and it cannot jump to point f' because of the transient; i.e., the signals change so rapidly that transient cannot die out. It will then return to point e again at nearly the same path; therefore, the hysteresis loop cannot be observed. By utilizing this phenomenon, the single neuron with excitatory self-connection for the specific parameter domains is applied to filter the signals having different frequencies; i.e., the neuro-module can perform as a low-pass filter (see more details in Sect. 5.1.1).

3.3 Evolutionary Algorithm

An evolutionary algorithm is used to develop the network's structure as well as to optimize its synaptic weights. This algorithm is inspired by the principles of natural evolution based on genetic variation and selection rules. There are many evolutionary algorithms which have been developed during the last 30 years; e.g., *Genetic Algorithms (GAs)* were introduced by J.H. Holland in 1970s [93], *Evolutionary Strategies (ESs)* were developed in the 1960s by I. Rechenberg [167] and H.P. Schwefel [185] and *Evolutionary Programming (EP)* was presented by L.J. Fogel et al. in the early 1960s [77, 78] and there are plenty of textbooks, e.g., [118, 145, 210], and conference series on this topic.

Here, *Evolution of Neural Systems by Stochastic Synthesis (ENS³)* [97] was employed as reference material for the production of a neural control unit as well as a neural preprocessing unit for the approach of artificial perception–action systems. It has the capability to develop size and connectivity structure as well as simultaneously optimize parameters of neuro-modules[2] like the synaptic weights and bias terms. It has been successfully applied to various optimization and control problems in robotics as well as signal processing, e.g., [74, 131, 203, 204, 217, 218, 224].

The ENS³ algorithm is an implementation of a variation–evaluation–selection cycle (Fig. 3.10) operating on a population[3] of n neuro-modules (p_i, $i = 1,..., n$).

A population p_i consists of two sets which are parents $P(t)$ and offspring $\hat{P}(t)$ where the parameter t denotes the generation of the population. It can be initialized ($t = 0$) by a population of an empty network consisting of only input and output neurons without any hidden neurons and connections or

---

[2] Here, neuro-modules are sometimes referred to as neural networks, neural modules or even just neural nets. However, all these terms mean exactly the same thing.

[3] One population can be determined as one problem or one fitness function.

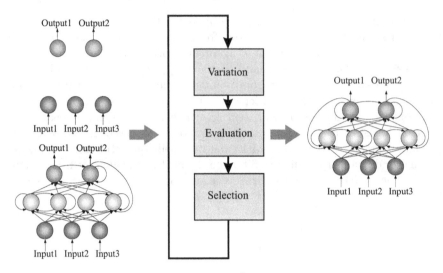

Fig. 3.10. The general function of the ENS[3] algorithm. The algorithm can start with the empty network (*upper picture on the left*) or the given network structure (*lower picture on the left*). The initial network is then presented to a variation-evaluation-selection process (cycle diagram shown in the *middle*). There is no formal stop criteria; i.e., it is repeated until the user manually stops the process, possibly if the reasonable network (picture on the *right*) is found

it can start with a given network structure (Fig. 3.10). However, there are two restrictions for using this evolutionary algorithm. One is that the transfer function of all neurons has to be the same. The other is that input neurons are solely used as a buffer; thus, no feedback connections to the input neurons are permitted. On the other hand, every kind of connection in hidden and output layers, e.g., self-connections, excitatory and inhibitory connections, is allowed.

Several operators in a variation–evaluation–selection cycle of the algorithm have to be considered for the evolutionary process. They can be formally represented as follows:

$$p(t+1) = S(E(V(R\ p(t)))), \tag{3.8}$$

where $p \in P(t) \cup \hat{P}(t)$ is a population of individuals and R, V, E and S are the reproduction, variation, evaluation and selection operators, respectively.

- The *reproduction* operator R creates a certain number of copies of each individual neuro-module from the parent group $P(t)$. The copies represent as the group of offspring $\hat{P}(t)$ in generation t. The number of copies is calculated by the selection operator S. This number is initially set to 1 for each module at the beginning ($t = 0$).

- The *variation* or *mutation* operator V is a stochastic operator and it is applied to offspring $\hat{P}(t)$ while the parents $P(t)$ are not allowed to change. It realizes the combinatorial optimization and real-valued parameter optimization. On the one hand, the combinatorial optimization refers to the fact that the number of hidden neurons and connections can be increased or decreased during the evolutionary process. It is determined by per-neuron and per-connection probabilities which are calculated according to a given probability and a random variable (0, 1). On the other hand, the real-valued parameter optimization refers to the variation of the bias and weight terms. It is calculated by using a Gaussian distributed random variable (0, 1).

- The *evaluation* operator E is defined in the term of a fitness function F that measures the performance or fitness value of each neuro-module. To keep the size of the evolved networks within limits, the fitness value takes into account the number of hidden neurons and connections; i.e., the desired number of neurons and connections can be negatively added to the fitness function by means of cost factors.

- The *selection* operator S is a stochastic operator. It selects which neuro-module from the group of the parent and offspring should be reproduced and passed to the next generation. This is achieved by taking into account the fitness value based on a ranking process and a Poisson distribution. A neuro-module becomes member of the parent set of the next generation if its number of offsprings is >0.

This evolutionary process has no formal stop criteria. Thus it is repeated until the interruption by the user takes place. This means that the user has to manually decide when the process has to stop by observing all essential parameters, e.g., fitness values.

The ENS[3] algorithm was integrated as a part of the *Integrated Structure Evolution Environment (ISEE)* [97]. It is a powerful software platform not only for the evolution of structures but also for nonlinear analysis of evolved structures and even for connecting different simulators as well as physical robot platforms. This ISEE platform combines three different components which are the evolution program *EvoSun*, the execution program *Hinton* and the simulators. The scheme of the ISEE is presented in Fig. 3.11.

At the beginning, individual neuro-modules are created in EvoSun (reproduction process) and then EvoSun sends the neuro-module information to Hinton for processing (evaluation process). Hinton executes one individual neuro-module at a time and communicates with a simulator. Two kinds of simulators, the *Yet Another Robot Simulator (YARS)* [2] and the *Data Reader*, are provided for the evolutionary process. Hinton has to be connected to one of them depending upon the desired task. If Hinton is connected to the YARS, then the motor and sensory data will be sent and received, respectively. In this case, the YARS is used to simulate walking machines together with

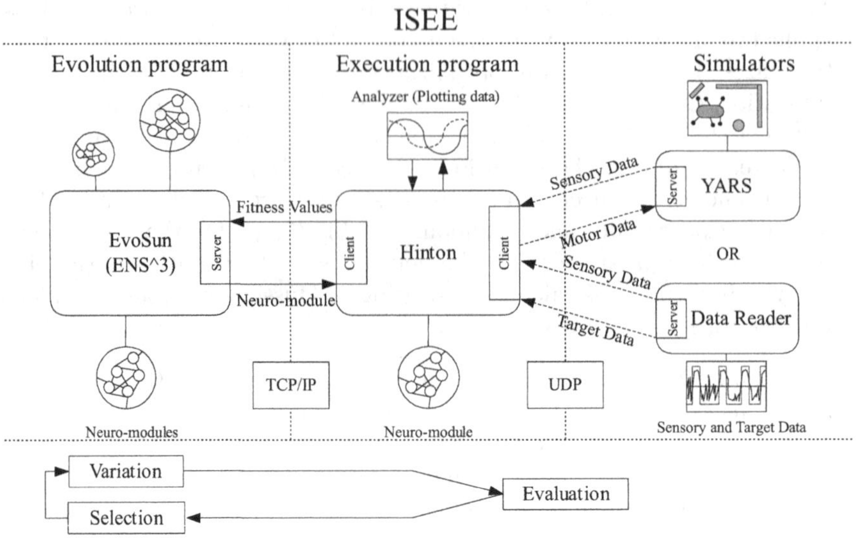

Fig. 3.11. The scheme of the evolutionary process with the ISEE modified from B. Mahn 2003, p. 32 [124] (see text for details)

their sensors (see also Sect. 4.2) in a virtual environment to test and optimize the neural control. The simulator processes a certain number of steps with the update frequency of 75 Hz, which is similar to the update frequency of the target system (a preprocessing of antenna-like sensors on a mobile processor). On the other hand, if Hinton is interfaced with the Data Reader, sensory data together with target data (or known as training data) will be received instead. It is used as the buffer of the sensory data and target data for evolving the neural preprocessing where the evolution task is the minimization of an error function. In this case, the executed neuro-module will be processed at an update frequency with respect to the update frequency of simulated or recorded sensor data. For example, it will be updated at 48 kHz if the sensor data is simulated or recorded through the sound card at a sampling rate of 48 kHz on a 1-GHz personal computer (PC). On the other hand, it will be updated at $\approx$ 2 kHz if the sensor data is recorded via a mobile system consisting of a personal digital assistant (PDA) and the Multi-Servo IO-Board (MBoard).

In both cases, the executed neuro-module is updated in accordance with the sensory data coming from the YARS or the Data Reader and a new output signal of the neuro-module is then calculated. The resulting output signal will be returned as motor data to the simulator, if Hinton is linked to the YARS. This updated execution–simulation process is continuously performed as long as a specified number of cycles is not fulfilled and a fitness value is constantly calculated according to a given fitness function. After that, the final fitness value of the executed neuro-module is sent back to EvoSun and a new individual neuro-module is again sent to Hinton for the execution and evaluation process until all individual neuro-modules of one generation are evaluated. EvoSun then selects a certain number of the neuro-modules (selection process) by taking into account the fitness value based on the ranking process. The selected neuro-modules, consisting of the parent and the offspring, are reproduced to the next generation and the offspring continues to the variation process afterwards. Each individual neuro-module of a new generation is again executed at Hinton and evaluated with the help of the simulator. This variation–evaluation–selection cycle is repeatedly run, until the evolutionary process is stopped by the user. During evolution the user is able to modify all essential evolution parameters: e.g., population size, variation probabilities, evaluation steps, cost factors of neurons and synapses, etc. Moreover, the user can on-line monitor the population parameters, evolution dynamics, properties of individuals, performance of individuals, and so on via EvoSun and the user can even analyze the resulting neuro-module via the analyzer tool implemented on Hinton.

3.4 Conclusion

Neural networks have a number of excellent properties; e.g., they can process information simultaneously, they can be adaptive, they exhibit dynamical behavior and they are even able to build a robot brain as an integration of different neuro-modules. Especially, for recurrent neuro-modules with specific parameter domains hysteresis effects can be observed. We can benefit from this dynamical effect in the preprocessing of sensory signals as well as in robot control. Therefore, artificial neural networks can be adequately employed for our work. Furthermore, the evolutionary algorithm ENS^3 was also presented, and it is realized by the principles of natural evolution in the form of variation, selection and evaluation rules. On the one hand, the algorithm permits every kind of connection in hidden and output layers. On the other hand, it maintains a network structure as small as possible with respect to the given fitness function. Thus, recurrent neuro-modules with a small network structure can be produced. Therefore, the ENS^3 is applied as a tool for developing and optimizing the neural preprocessing and control to achieve artificial perception–action systems.

4

Physical Sensors and Walking Machine Platforms

This chapter describes the development of the physical components that lead to the artificial perception–action systems. It begins with the descriptions of different physical sensors which are used to sense environmental information, followed by the details of the walking machines simulated in a physical simulation environment as well as the robots we have built.

Inspired by the function of the hairs of a spider for sound detection and a scorpion for tactile sensing, an artificial auditory–tactile sensor in analogy to these sensory hairs is introduced. In addition, the set-up of a so-called stereo auditory sensor together with its electronic circuit for sound tropism approach is also presented. We then discuss the use of physical infrared sensors as a functionally equivalent antenna model for detecting obstacles. Finally, the design and the construction of biologically inspired four- and six-legged walking machines with different morphologies as well as their physical simulators are presented.

4.1 Physical Sensors

To generate the different reactive behaviors of the walking machines in accordance with an environmental condition, sensory information is required. Three physical sensor systems for providing the signals to trigger several behaviors were implemented and tested on physical walking machines.

4.1.1 An Artificial Auditory–Tactile Sensor

The wandering spider *Cupiennius salei* preys on a flying insect by using the special sensory hair (Trichobothria) on its limbs which is also sensitive to the auditory cues in a low-frequency range described in Sect. 2.1. Differently, the scorpion *Pandinus cavimanus* uses its hairs as tactile sensors to perform several tasks, e.g., an obstacle avoidance task (see also Sect. 2.1). Analogs of these auditory and tactile hair sensor systems of spiders and scorpions can be useful in providing environmental information for a sensor-driven control system in wheeled robots as well as in walking machines.

There exist implementations of the tactile sensors [92, 105, 106] and the auditory sensors [95] on real robots but roboticists have not yet implemented these two sensor functions into one sensor system. However, M. Lungarella et al. [122] and H. Yokoi et al. [222] introduced an artificial whisker sensor with a real mouse whisker attached, hair-like, to a capacitor microphone. In the works of M. Fend et al. [67, 68], the whisker sensors were applied for an obstacle avoidance task and texture discrimination while the use of the sensors for sound detection was not mentioned.

Here, the whisker sensor is applied for the auditory–tactile application [127]. It will enable autonomous mobile robots as well as walking machines to move around for indoor applications. The sensor shall protect a robot's body and especially the legs of walking machines from colliding with obstacles, like chair or desk legs. In addition, with the implementation of the sound tropism, the robot will also be able to navigate. A so-called auditory–tactile sensor consists of a mini-microphone (0.6-cm diameter) built in an integrated amplifier circuit, a root (a small rubber wire) and a whisker-shaped material taken from a whisker of a real mouse (4.0-cm long). The sensor and its components are shown in Fig. 4.1.

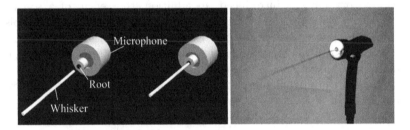

Fig. 4.1. The auditory–tactile sensor consists of a whisker of a real mouse, a rubber root and a capacitor microphone built in an integrated amplifier circuit. *Left*: Assembly parts of a sensor. *Right*: The real sensor built in an amplifier circuit

In order to build this sensor, the mouse whisker was inserted into a root which was glued onto the diaphragm of a microphone. The physical force of the whisker vibrates the diaphragm of the capacitor microphone, which results in a voltage signal. The signal is amplified via the integrated amplifier circuit on the mini-microphone. The maximum output voltage with respect to the given input signals, e.g., a sine wave signal, is around 1.8 peak volt AC. To record the signal via a line-in port of a PC sound card, it has to be scaled in the range of a maximum output voltage at around 0.5 peak volt AC. It was done by using a potentiometer which functions as a variable voltage divider. The scaled output signal is then digitized on the sound card at a sampling rate of 48 kHz for the purpose of monitoring and feeding it into the neural preprocessor afterwards. The basic scheme of the sensor system is shown in Fig. 4.2.

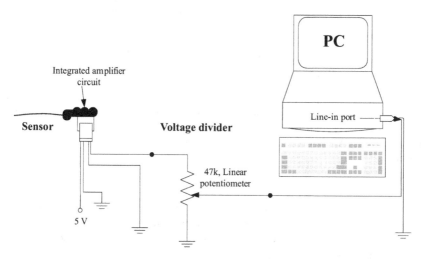

Fig. 4.2. The basic scheme of the auditory–tactile sensor system. The detected signal is first amplified via an integrated amplifier circuit of the mini-microphone, then the amplitude of an amplified signal is reduced by a variable voltage divider. Eventually, the scaled signal goes into the line-in port for digitizing and then is fed into the neural preprocessor

By applying this sensor system to obtain tactile and auditory signals, one should keep in mind that the tactile signal requires a high sampling rate of an analog to digital converter (ADC), e.g., 48 kHz, while the auditory signal depending on a used frequency can be digitized at a lower sampling rate. The response of the sensor to an auditory signal and a tactile signal recorded via the line-in port is exemplified in Fig. 4.3.

In comparison to the auditory and tactile signals (Fig. 4.3), the tactile signal has a vibrating waveform with slightly higher frequency while the auditory signal has a sine waveform with a lower frequency. Such different signal properties are crucial to seek signal processing by using neural network and evolutionary approaches (described in Chap. 5).

4.1.2 A Stereo Auditory Sensor

To perform a sound tropism which is inspired by the prey capture behavior of the spider *Cupiennius salei* (see also Chap. 2), a so-called stereo auditory sensor is employed. The sensor together with its signal processing (described

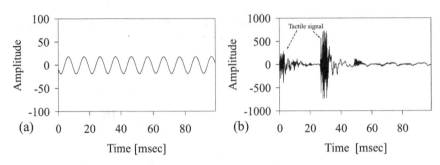

Fig. 4.3. (a) The response of the auditory–tactile sensor to an auditory signal at a frequency of 100 Hz which is generated by a loudspeaker. (b) The response of the sensor to a tactile signal which is generated by sweeping the sensor over an object back and forth. All figures have the same scale in the x-axis while the y-axis is different

in Chap. 5) will enable a walking machine to detect sound[1] and discern the direction of the source. The processed sensor signals can then control an autonomous mobile robot to move in the direction of the sound source and make an approach to it in a real environment.

There are several examples of robot experiments that use an auditory sensor–in a form of a microphone–for different purposes. Most researchers use an array of four or more microphones to perform auditory source localization [3, 195, 206, 209]. Such a system is too expensive to compute a signal processor, too complex and also too energy consuming, despite the fact that it can detect the signals in three-dimensional space and precisely localize the source. There are other examples, for instance, the SAIL robot uses the microphone for online learning of verbal commands [226] and a humanoid robot called ROBITA uses two microphones to follow a conversation between two persons [133]. The behavior generated by auditory signals is also studied in [95, 169, 215]. They used two miniature microphones allowing the robot to detect and move toward a simulated male cricket song—4.8 kHz [121, 134].

The above-stated research study shows that the use of a microphone can achieve several tasks and even two microphones are adequate to perform sound source localization in two-dimensional space. Thus, in this book, the stereo auditory sensor system was built from two miniature microphones with a 0.6-cm diameter (for the left and the right detections in two-dimensional space), a support circuitry and the MBoard. The system was suitably implemented on a four-legged walking machine.

Concerning time delay of arrival (TDOA) [54, 138] of the sound coming from the two microphones (later called the stereo auditory sensor), the microphones were installed on the (moving) fore left and rear right legs of the walk-

---

[1] Here, sound having a sine waveform at a frequency of 200 Hz is used for a sound tropism approach. The sound with this property is later called an auditory signal.

ing machine. Consequently, they can scan the auditory signals in the wider angle because they are moving with the legs. The locations where the stereo auditory sensor (the left and right microphones) were installed are shown in Fig. 4.4.

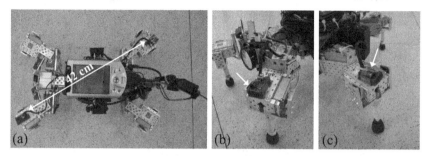

Fig. 4.4. (a) The distance between the microphones of the stereo auditory sensor is equal to 42 cm. (b) The real sensor built in a preamplifier circuit was installed on the left foreleg of the walking machine. (c) The sensor was installed on the right hind leg of the walking machine

The auditory signals are initially amplified via the microphones' integrated amplifier circuit, and then scaled to the range between 0 and 5 volts by a support circuitry. Afterwards, they are digitized via ADC channels of the MBoard at a sampling rate of up to 5.7 kHz. To obtain the sensor data, the MBoard can be interfaced with either a PC or a PDA via a serial (RS232) port. The basic scheme of the sensor system is shown in Fig. 4.5.

According to the dimension of the walking machine and the distance between the fore left and the rear right microphones, the maximum time delay between the left and the right is equivalent to one-fourth of the wavelength of the frequency—200 Hz. The response of the sensor to the auditory signals recorded via the MBoard and displayed on a 1-GHz PC is exemplified in Fig. 4.6.

As shown in Fig. 4.6, the desirable occurrence of time delay between the left and the right microphones in accordance with the location of the sound source will be used to seek signal processing to generate a sound tropism. In addition, the amplitude of the signal will also be used to estimate the distance between the walking machine and the source where high amplitude indicates to approach the source and vice versa (see more details in Chap. 5).

4.1.3 Antenna-like Sensors

In order to achieve versatile artificial perception–action systems, the walking machine should not only have sound tropism, but it should also perform other

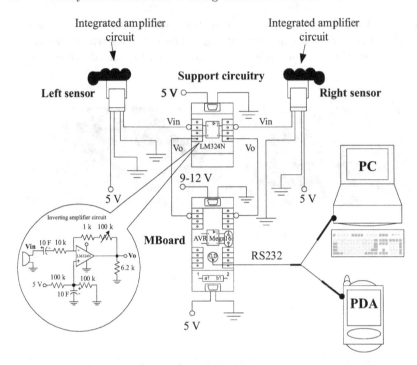

Fig. 4.5. The basic scheme of the stereo auditory sensor system. The detected signals coming from the left and right microphones are initially amplified via the integrated amplifier circuit of the microphone. Then, amplified signals are scaled to a range between 0 and 5 volts through the support circuitry. After that, the MBoard digitizes the scaled output voltages to a 7-bit value, where 0 represents silence and 128 represents maximum volume. Eventually, the digital signals from the MBoard are displayed on a PC or a PDA via an RS232 interface at a transfer rate of 57.6 kbits/s

behaviors like an animal, e.g., wandering and avoiding objects or even escaping from a deadlock situation.

Therefore, additional sensors which can detect obstacles are required. Inspired by an insect antenna (cf. Chap. 2), our physical sensors were modeled using the infrared (IR) sensors. An IR sensor has a lot in common with an insect antenna. Although an IR sensor acts differently from an insect antenna, by measuring the brightness of the IR light reflected by objects, the resulting measurement is the same. It is a well-known fact in robotics, that using IR sensors instead of antennas is a simplification as well as a solution with low power consumption. Most researchers use the sensors in a mobile robot as well as in a walking machine for obstacle avoidance [72, 74, 128] or even wall following [46].

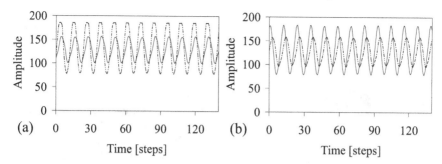

Fig. 4.6. (a) The sound source is close to the fore left microphone. This results in the signal coming from the fore left microphone (*dashed line*) having high amplitude and it is followed by the signal coming from the rear right microphone (*solid line*) with a delay, while the reverse case is presented in (**b**). All figures have the same scale in the *x*-axis and the *y*-axis

In this book, three types of the IR sensor, later called "antenna-like sensors", were chosen to detect obstacles at distances of 4–30 cm, 10–80 cm and 20–150 cm. The antenna-like sensors were implemented and tested on two different walking machines (four-legged and six-legged walking machines). Two antenna-like sensors which can detect obstacles at a distance of 10–80 cm were installed on the (moving) forehead of the four-legged walking machine AMOS-WD02.[2] They make an angle of approximately 25 degrees with respect to the horizontal body axis of the walking machine. The angle was manually adjusted for optimal operation. Consequently, the walking machine is able to detect obstacles on the fore left and right of its body (Fig. 4.7).

As a result of the structure of the four-legged walking machine, its head, where the sensors were implemented, can vertically turn left and right with respect to the walking pattern by activating the backbone joint. Consequently, the sensors can also scan obstacles in a wider angle. In other words, they perform like an active antenna scanning an obstacle in two-dimensional space (Fig. 4.8).

Normally, two antenna-like sensors on its left and right foreheads are sufficient to perform an obstacle avoidance. However, to prevent the legs of the walking machines from hitting obstacles, like chair or desk legs, more sensors are needed and they can be installed on the (moving) legs.

Here, the six sensors were implemented on the six-legged walking machine AMOS-WD06. Two of them, which can detect the obstacle at a long distance of 20–150 cm, were fixated at the forehead while the rest of them, operating at a shorter distance 4–30 cm, were fixated at the two forelegs and two middle legs. The configuration of the sensors on the AMOS-WD06 and the idealized field of the sensors are presented in Fig. 4.9.

---

[2] Advanced MObility Sensor driven-Walking Device.

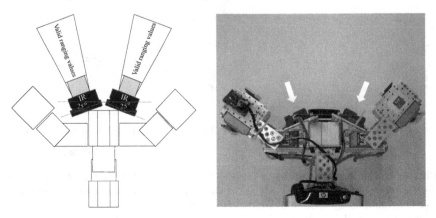

Fig. 4.7. The antenna-like sensors implemented on the forehead of the four-legged walking machine. *Left*: The outline of the sensors from a top view. *Right*: The real sensors fixated on the forehead of the physical four-legged walking machine AMOS-WD02 (*arrows*)

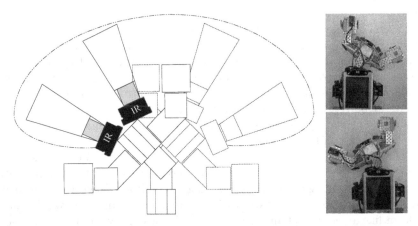

Fig. 4.8. The idealized field of the antenna-like sensors when the backbone joint of the walking machine is activated. *Left*: The outline of the idealized field where the sensors can scan obstacles (*dashed curve*). *Right*: The visualization of the sensors moving with the head of the walking machine when the backbone joint turns right (*upper picture*) and left (*lower picture*)

As shown in Fig. 4.9, one pair of the forehead sensors performs like a passive antenna detecting obstacles in front of the walking machine, while the other two pairs installed on the (moving) legs perform like active antennas because they move along the legs. Therefore, these (active) sensors can scan the obstacle in three-dimensional space; i.e., they move forward and backward

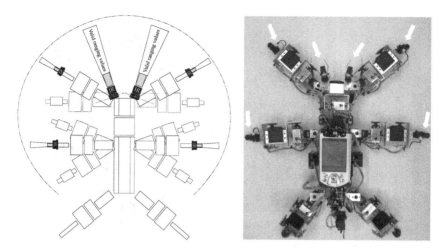

Fig. 4.9. *Left*: The visualization of the locations where the sensors are implemented and the idealized field of the sensors protecting the walking machine from crashing into obstacles (*dashed line* around the walking machine). *Right*: The six sensors on the physical walking machine (*arrows*)

in parallel to the ground (Fig. 4.9) and they also move up and down in a vertical direction (Fig. 4.10).

To obtain the sensory data for controlling the behavior of the walking machine, all sensors were interfaced and digitized via the ADC channels of the MBoard at the sampling rate of up to 5.7 kHz. Subsequently, the digital signals are sent to either a PC or a PDA through an RS232 interface at a transfer rate of 57.6 kbits/s for the purpose of monitoring and feeding the data afterwards into the preprocessing network. The basic scheme of the sensor system is shown in Fig. 4.11.

The example of the sensor signals responding to a presented object is shown in Fig. 4.12. As shown in Fig. 4.12, the sensor signals have some noise resulting in uneven signals, and this may lead to difficulties in controlling the behavior of the walking machines. Therefore, the preprocessing of these sensor signals, described in Sect. 5.1.3, is required to eliminate the unwanted sensory noise and to trigger the obstacle avoidance behavior of the walking machines.

4.2 Walking Machine Platforms

To demonstrate reactive behaviors and to experiment with neural controllers, a mobile robot platform is required, and it would be desirable that it has a morphology like a walking animal (cf. Sect. 2.2). There are several examples for the construction of the four- and six-legged walking machines. Most of them

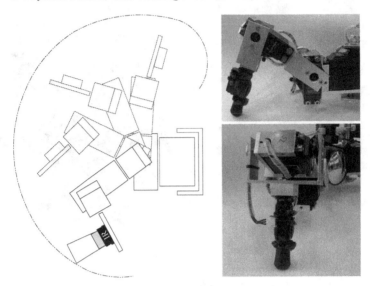

Fig. 4.10. *Left*: The idealized field of the antenna-like sensor on the right foreleg of the six-legged walking machine (*dashed curve*) with the remaining sensors operating on the other legs. The sensor moves in a vertical direction when the basal and distal joints are activated. *Right*: The physical sensor on the right foreleg

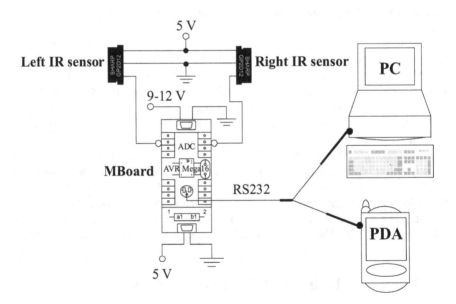

Fig. 4.11. The basic scheme of the antenna-like sensor system. Here, two sensors are presented. They are connected to the ADC channels of the MBoard. The digital signals from the MBoard will be displayed or analyzed on a PC or a PDA via an RS232 interface

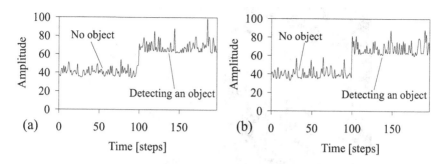

Fig. 4.12. The sensory signals coming from the left forehead (**a**) and the right forehead (**b**) of the four-legged walking machine

were designed to have a trunk without a backbone joint, such as *Lobster* [12], *Sprawlita* [47], *Tarry I, II* [80], *LAURON series robots* [81], *ARAMIES* [91], *Warp1* [102], *TITAN series robots* [107], *AirBug* [109], *Jumping Quadruped* [110], *TUM* [162], *Cockroach series robots* [164] and *Scorpion* [191]. However, some of these examples gain the benefits of having different configurations which promote stability and flexibility of locomotion while maintaining animal characteristics *BISAM* [28], *Robo-Salamander* [34], *Hexamos* [178], *MechaRoach II* [216] and *Hyperion* [223].

Thus, the four- and six-legged walking machines were constructed with different morphologies analogous to the principal structures of a salamander and a cockroach, respectively. Their structures were initially designed and visualized in 3D models before assembling the physical components in the final stage. Furthermore, a physical simulation was used to create the walking machines in the virtual world to test and experiment with neural controllers before downloading them into the real-world walking machines.

4.2.1 The Four-Legged Walking Machine AMOS-WD02

The AMOS-WD02 [125] consists of four identical legs. Each leg has two joints (two degrees of freedom (DOF)), which are a minimum requirement to obtain the locomotion of a walking machine and which follow the basic principle of movement of a salamander leg (cf. Sect. 2.2). The upper joint of the legs, called the thoracic joint, can move the leg forward (protraction) and backward (retraction), and the lower one, called the basal joint, can move it up (elevation) and down (depression) [12] (Fig. 4.14).

The length of the levers which are attached to the basal joints is proportional to the dimension of the machine. They are kept short to avoid greater torque in the actuators [163]. The configuration of the leg, built from a construction kit [33], is shown in Fig. 4.13.

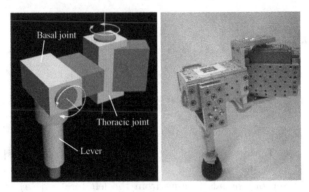

Fig. 4.13. The leg with two DOF. *Left*: The 3D model of the leg. *Right*: The physical leg of the AMOS-WD02

Inspired by vertebrate morphology of the salamander's trunk and its motion (described in Sect. 2.2), the robot was constructed with a backbone joint which can rotate around a vertical axis. It facilitates a more flexible and faster motion.[3] The backbone joint is also used to connect the trunk, where two hind legs are attached, with the head, where two forelegs are installed. The trunk and the head are formed with the maximum symmetry to keep the machine balanced for stability while walking. They are also designed to be as narrow as possible to ensure optimal torque from the supporting legs to the center line of the trunk. The construction of the walking machine together with the working space of the legs and the active backbone joint is shown in Fig. 4.14. The detail of the dimension is presented in Appendix A.

Moreover, a tail with two DOF rotating in the horizontal (y-axis) and vertical (z-axis) axes was implemented on the back of the trunk. In fact, this actively moveable tail, which can be manually controlled, is used only to install a mini wireless camera for monitoring the environment while the machine is walking. However, the tail also gives the walking machine a more animal-like appearance, e.g., in analogy to a scorpion's tail with its sting [4] (Fig. 4.15).

All leg joints are driven by analog modelcraft servomotors producing a torque between 70 and 90 Ncm. The backbone joint is driven by a digital servomotor with a torque between 200 and 220 Ncm. For the tail joints, micro-analog servomotors with a torque around 20 Ncm were selected. The height of the walking machine is 14 cm without its tail, and the weight of the fully equipped machine (including 11 servomotors, all electronic components, battery packs and a mobile processor) is approximately 3.3 kg. In addition, this machine has two antenna-like sensors and two auditory sensors to perform

---

[3] A walking speed is approximately 12.7 cm/s when the backbone joint is inactivated, while it is approximately 16.3 cm/s with the activation of the backbone joint in accordance with the walking pattern. The measurements were done with the walking frequency of the machine at 0.8 Hz.

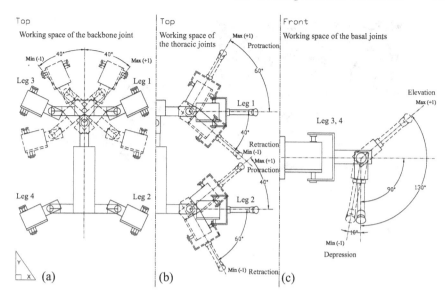

Fig. 4.14. (a) The angle range of the backbone joint (*top view*). (b) The angle ranges of all thoracic joints on the right side of the walking machine with the left side being symmetric (*top view*). (c) The angle range of the basal joint of the left foreleg with the remaining legs having the same angle ranges (*front view*)

Fig. 4.15. *Left*: A scorpion's tail with a sting (modified from S.R. Petersen 2005 [160] with permission). Middle: The tail of the four-legged walking machine AMOS-WD02. *Right*: The tail of the six-legged walking machine AMOS-WD06. The two DOF tail is constructed in an abstract form of a scorpion's tail. It is mainly used to install the camera

different reactive behaviors; e.g., an obstacle avoidance and a sound tropism, respectively. The 3D model of the walking machine and the real walking machine are shown in Fig. 4.16.

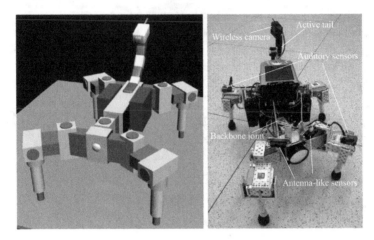

Fig. 4.16. The four-legged walking machine AMOS-WD02. *Left*: The 3D model of the walking machine. *Right*: The real walking machine

All in all, the AMOS-WD02 has 11 active DOF, 4 sensors and 1 wireless camera (for more details of the AMOS-WD02, see Appendix A). Therefore, it can serve as a reasonably complex platform for experiments concerning the function of the neural perception–action systems.

However, to test the neural controller and to observe the resulting behavior of the walking machine (e.g., obstacle avoidance), they were first simulated in the physical simulation environment YARS (cf. Sect. 3.3). The simulator, developed at the Fraunhofer Institute in Sankt Augustin, is based on Open Dynamics Engine (ODE) [189]. It provides a defined set of geometries, joints, motors and sensors which is adequate to create the four-legged walking machine AMOS-WD02 with IR sensors in a virtual environment with obstacles (Fig. 4.17).

The YARS enables first implementation which is precise enough to reproduce the behavior of the physical walking machine with sufficient quality. This simulation environment is also connected to the ISEE, which is a software platform for developing neural controllers (described in Chap. 3).

In the final stage, a neural controller which is developed after the test on the simulator is then applied to the physical walking machine to demonstrate the behavior in the real environment. The neural controller is programmed into a mobile processor (a PDA). The PDA is interfaced with the MBoard, which digitizes sensory signals and generates a pulse width modulation (PWM) signal at a period of 20 ms, to command the servomotors. The communication between the PDA and the MBoard is accomplished via an RS232 interface at 57.6 kbits/s.

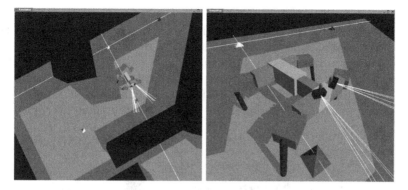

Fig. 4.17. Different views of the simulated walking machine in its environment. The properties of all simulated components are defined with respect to the physical properties of the real walking machine, e.g., weight, dimension, motor torque and so on. The simulated walking machine consists of body parts (head, backbone joint, trunk and limbs), servomotors and IR sensors, while the auditory sensor was not available in the simulation

4.2.2 The Six-Legged Walking Machine AMOS-WD06

The AMOS-WD06 [126] consists of six identical legs, and each leg has three joints (three DOF), which is somewhat similar to a cockroach leg. A thoracic joint has similar functionality to the thoracic joint of the AMOS-WD02, while another two joints, the basal and distal joints, are used for lifting (elevation) and lowering (depression) and for extension and flexion of the leg [12]. The levers which are attached to distal joints were built in the same manner as the levers of the AMOS-WD02. The configuration of the leg is shown in Fig. 4.18.

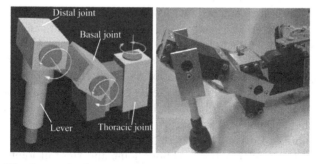

Fig. 4.18. The leg with three DOF. *Left*: The 3D model of the leg. *Right*: The physical leg of the AMOS-WD06

This leg configuration provides the machine with the ability to perform omnidirectional walking; i.e., the machine can walk forwards, backwards, laterally and turn with different radii. Additionally, the machine can also perform a diagonal forward or backward motion to the left or the right by activating the forward or backward motion together with the lateral left or right motion. The high mobility of the legs enables the walking machine to walk over an obstacle, stand in an upside-down position or even climb over obstacles (Fig. 4.19).

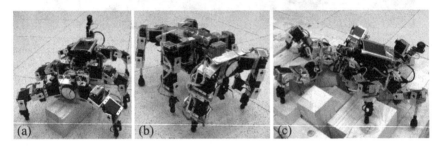

Fig. 4.19. The walking machine AMOS-WD06 walking over an obstacle with the maximum height of 7 cm (**a**), standing in an upside-down position (**b**) and climbing position on obstacles which is enabled by the active backbone joint (**c**)

Inspired by the invertebrate morphology of the American cockroach's trunk and its motion (described in Sect. 2.2), a backbone joint which can rotate in a horizontal axis was constructed. It is designed to operate like a cockroach while the machine is climbing over obstacles (Fig. 4.19c). However, this active backbone joint will be fixed under normal walking conditions of the machine. Mainly, it is used to connect the trunk, where two middle legs and two hind legs are attached, with the head, where two forelegs are installed. The trunk and the head were designed with the same concepts as the AMOS-WD02 described above. The construction of the AMOS-WD06 together with the working space of the legs and the active backbone joint is shown in Fig. 4.20 (see also in Appendix A).

Similar to the AMOS-WD02, one (active) tail with the same configuration was also implemented on the back of the trunk (Fig. 4.15). It has a similar function as the AMOS-WD02's tail.

All leg joints are driven by analog servomotors producing a torque between 80 and 100 Ncm. For the backbone joint and the tail joints, the same motors which were used on the AMOS-WD02 were employed. The height of the walking machine is 12 cm without its tail, and the weight of the fully equipped robot (including 21 servomotors, all electronic components, battery packs and a mobile processor) is approximately 4.2 kg. Like the AMOS-WD02, a mini wireless camera with a built-in microphone was installed on the tail

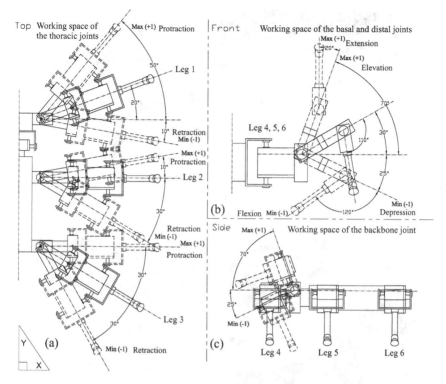

Fig. 4.20. (a) The angle ranges of all thoracic joints on the right side of the walking machine with the left side being symmetric (*top view*). (b) The angle ranges of the basal and distal joints of the left foreleg with the remaining legs having the same angle ranges (*front view*). (c) The angle range of the backbone joint (*side view*)

for monitoring and observing the environment while walking. In addition, the walking machine has six antenna-like sensors to help detect obstacles and one upside-down detector which is implemented beside the trunk of the machine. The 3D model of the walking machine and the real machine are shown in Fig. 4.21.

All in all, the AMOS-WD06 has 21 active DOF, 7 sensors and 1 wireless camera (for more detail of the AMOS-WD06, see Appendix A); thus it can also serve as a testing platform like the AMOS-WD02. The AMOS-WD06 was also simulated by the YARS with the same virtual environment and the same purpose as described above. The basic features of the simulated walking machine are closely coupled to the physical walking machine, e.g., weight, dimension, motor torque and so on. It consists of body parts (head, backbone joint, trunk and limbs), servomotors, IR sensors and an additional tail. The simulated walking machine with its virtual environment is shown in Fig. 4.22.

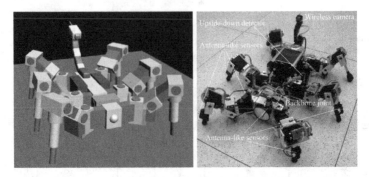

Fig. 4.21. The six-legged walking machine AMOS-WD06. *Left*: The 3D model of the walking machine. *Right*: The real walking machine

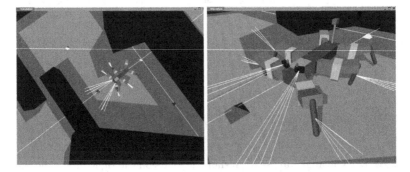

Fig. 4.22. Different views of the simulated walking machine in its environment

The final neural controller will also be implemented on the physical walking machine for testing its behavior in a physical environment. Again, the controller is programmed on the same mobile processor system with the same update frequency as the AMOS-WD02.

4.3 Conclusion

We used three types of physical sensor systems: an auditory–tactile sensor, a stereo auditory sensor and antenna-like sensors. The auditory–tactile sensor, which was inspired by the function of the hairs of a scorpion and a spider, can be used for tactile sensing as well as sound detection. Using the stereo auditory sensor in analogy to the hairs of the spider, the sound can be detected and the direction of the incoming sound can also be distinguished by determining the TDOA from the left and right auditory sensors. The antenna-like sensors are used to detect impediments as well as to protect the legs of the walking machine from colliding with obstacles.

Two walking machines with different morphologies were built with physical components. They were also simulated in a physical simulation environment with the intention to develop and test neural controllers before implementation in the real-world walking machines. Thus, the walking machines together with the sensor systems can serve as hardware platforms for experiments with neural controllers and for artificial perception–action systems.

5

Artificial Perception–Action Systems

Where Chap. 2 investigated the biologically inspired perception–action systems, this chapter focuses on applying the principles of the biological domain to create artificial perception–action systems. First, several preprocessing units of different types of sensory signals are presented. They are used to filter and recognize the corresponding sensory signals and they can be described as perception parts. Second, the neural control of the four- and six-legged walking machines, which generates and controls the locomotion of the machines, is described. Third, the combination of the neural preprocessing and the neural control is explained. It gives rise to the ability of controlling reactive behaviors such as obstacle avoidance and sound tropism. Finally, both behavior controls are merged under a so-called behavior fusion controller by applying a sensor fusion technique to give a versatile perception–action system.

5.1 Neural Preprocessing of Sensory Signals

We shall now present three different types of neural preprocessing modules which use the dynamic properties of recurrent neural networks (as described in Chap. 3). The first module is a so-called *auditory signal processor* which is used to preprocess the auditory signals detected by means of a stereo auditory sensor or auditory–tactile sensors. It consists of two subordinate networks, one for filtering auditory signals to detect the low-frequency sound, and the other to distinguish the direction of detected signals between the right and the left. The second module is known as the *tactile signal processor* and it has the capability to recognize the tactile information coming from an auditory–tactile sensor. The last module does the *preprocessing of antenna-like sensor data*, which can eliminate the sensory noise, and its outputs are used to control the walking behavior of the machines for avoiding obstacles or even escaping from a corner.

5.1.1 Auditory Signal Processing

Inspired by the function of the sensory hair of the spider (cf. Chap. 2), the auditory signal processing is studied. The function of the auditory signal processing is similar to the described sensory and sensing systems. It enables the walking machine(s) to recognize low-frequency sound and to distinguish the auditory signals coming from the left or the right. In order to create such signal processing, first a simple network that acts as a low-pass filter is investigated [127]. Subsequently, the other network which will help the machine(s) to discern the direction of a sound source is constructed. At the end, the integration of both networks leads to the complete auditory signal processing network. This effective network is then applied to preprocessing the signals of the stereo auditory sensor or the auditory–tactile sensors.

A Low-Pass Filter for Auditory Signals

In order to have a network which can detect low-frequency sound, an artificial neural network together with an evolutionary algorithm is employed. Also, an input signal of sine shape which is a mixture of 100 Hz and 1000 Hz is simulated on a 1-GHz PC with an update frequency of 48 kHz. The input signal is mapped to a range between −1 and +1, and then it is buffered into the simulator called *Data Reader* (described in Sect. 3.3) for the purpose of feeding the data to evolve or test the network. To keep the problem simple an ideal noise-free signal with constant amplitude (Fig. 5.1a) is used at the beginning. If a network, which is found, can distinguish between low-frequency (100 Hz) and high-frequency (1000 Hz) sounds, the next step of the experiment is to apply the signal with varying amplitudes (Fig. 5.1b) which is recorded via a physical auditory–tactile sensor. The recorded signal is digitized through the line-in port of a sound card at a sampling rate of 48 kHz on a 1-GHz PC.

To design the neural preprocessing structure, a single model neuron configured as a hysteresis element [151] is utilized; i.e., the network consists of an input neuron and a neuron with a positive self-connection corresponding to a dynamical neural Schmitt Trigger [96] (Fig. 5.2a). The network is constructed, experimented and analyzed through the ISEE connecting with the Data Reader. This is the software platform for developing neural controllers, and it is implemented on a 1-GHz PC (described in more detail in Sect. 3.3). In this case, the network is updated at a frequency of 48 kHz. Applying the results from [96], the weight ($W_1 = 1$) from the input unit to the output unit and the bias term ($B = -0.1$) are fixed while the self-connection weight W_2 of the output unit is varied from 0 to 2.5. The ideal noise-free signal with constant amplitude (Fig. 5.1a) is given to the network. For $W_2 = 2.45$ the network suppresses high-frequency sound of 1000 Hz, while low-frequency sound of 100 Hz passes through it. The resulting network, called the "simple auditory network", is shown in Fig. 5.2.

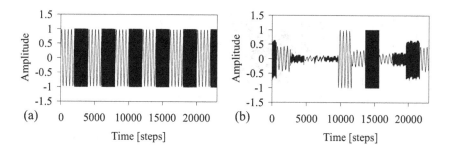

Fig. 5.1. The input signal of sine shape mixed between 100 Hz and 1000 Hz. (**a**) The simulated noise-free signal with constant amplitude. (**b**) The noisy signal with varying amplitudes recorded via the physical sensor. Both signals are updated at a frequency of 48 kHz

In addition, the resulting network is tested with varying frequencies of an input signal from 100 Hz to 1000 Hz (Fig. 5.3a). The output signal shows that it can detect the signal at a frequency up to approximately 300 Hz (see *dashed frame* in Fig. 5.3b), where this frequency is defined as the cutoff frequency of the network which is represented in Fig. 5.2b.

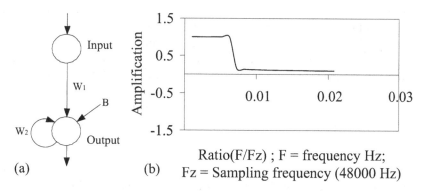

Fig. 5.2. (**a**) The simple auditory network realizing a low-pass filter; parameters are $W_1 = 1$, $W_2 = 2.45$ and $B = -0.1$. (**b**) The characteristic curve of this network with its cutoff frequency at approximately 300 Hz

From the result, it can be observed that this simple auditory network with its specific parameters has the property of a low-pass filter. By varying a weight W_2 of the self-connection of the output unit, one observes a splitting of the output signal, due to the hysteresis effect, which is different at various

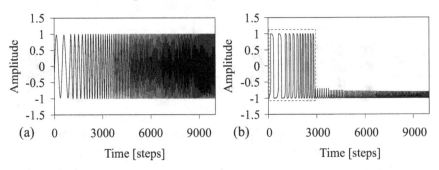

Fig. 5.3. (a) The varying frequencies of an input signal from 100 Hz to 1000 Hz. (b) The output signal of the network. The *dashed frame* is the frequency range (from 100 Hz to approximately 300 Hz) in which the network can detect the auditory signal

frequencies. This suggests that the hysteresis domain of a single neuron with self-connection [151] can play an important role for the filtering of signals.

To visualize this phenomenon, output versus input for low- and high-frequency signals are plotted in Fig. 5.4, and the different "hysteresis effects" can be compared in accordance with the different strengths of the self-coupling. Figure 5.4 shows that the hysteresis effect for high-frequency sound has already occurred for $W_2 = 0.25$, although it cannot yet be observed for low-frequency sound. If W_2 is increased up to $W_2 = 2.45$, high-frequency sound will almost be suppressed (a small amplitude of the output signal), whereas the hysteresis effect for low-frequency sound switches the amplitude between almost saturation values (between approximately -1 and $+1$). Increasing the strength of the self-connection up to $W_2 = 2.50$ low-frequency sound is also suppressed.

As the bias term defines the base activity of the neuron, the amplitude of high-frequency output is compensated, and it oscillates with small amplitude between -0.804 and -0.998. Eventually it will never rise above 0 again. In this situation, we suggest a low-pass filter function for a configuration with this specific bias (-0.1) and weight ($W_2 = 2.45$) (cf. Sect. 3.2). The neural network behaves as a low-pass filter because the output amplitude of high-frequency sound stays around -0.9 while the output amplitude of low-frequency sound remains oscillating between -0.997 and 0.998.

Having established the single neuron to act as a low-pass filter for noise-free signals of constant amplitude, the following step is to derive a network, which behaves like a robust low-pass filter and which is capable of recognizing low-frequency sound in the real environment. The input signal presented to the network is recorded through the physical auditory–tactile sensor and digitized at a sampling rate of 48 kHz. Then it is again mapped to a range between -1 and $+1$ and preserved into the Data Reader (Fig. 5.1b). The simple auditory network is now improved by adding one self-connected hidden unit, and by

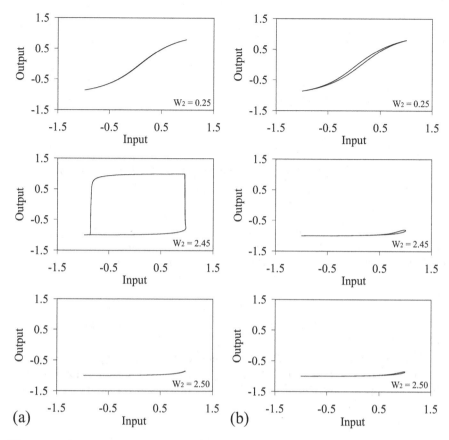

Fig. 5.4. Comparison of the "hysteresis effects" between input and output signals of high- and low-frequency sounds for $W_2 = 0.25$, 2.45 and 2.50, respectively. (**a**) Low-frequency sound (100 Hz); (**b**) High-frequency sound (1000 Hz)

manually adjusting the weights via the ISEE. With specific parameters, the network behaves like a robust low-pass filter; i.e., it can detect the noisy low-frequency sound. The final result, an advanced auditory network,[1] is shown in Fig. 5.5.

One should remark that the network can recognize the input signal only if the amplitude of an input signal is higher than the threshold, here 0.5. For this reason, it is also relevant to the sensing system of the spider because it can detect the signal of its prey at a close distance (see also Sect. 2.1.2),

---

[1] The network is named the advanced auditory network because of its uncomplicated neural structure and its performance that can detect the noisy low-frequency signal with varying amplitudes.

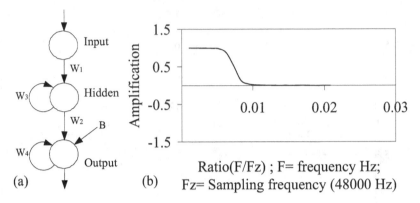

(a)   (b)   Ratio(F/Fz) ; F= frequency Hz;
Fz= Sampling frequency (48000 Hz)

Fig. 5.5. (a) The principal advanced auditory network, performing as a low-pass filter of the noisy signal with varying amplitudes. The bias term B is equal to -6.7 and all weights are positive, $W_1 = 0.01$, $W_2 = 32$, $W_3 = 1$ and $W_4 = 0.27$. (b) The characteristic curve of this network with its cutoff frequency at approximately 400 Hz

meaning that the amplitude of a detected signal should also be higher than the threshold.

To consider the characteristics of the network, an input signal with varying frequencies from 100 Hz to 1000 Hz having constant amplitude is presented to the network (Fig. 5.6a). The output signal is plotted in Fig. 5.6b with respect to the given input.

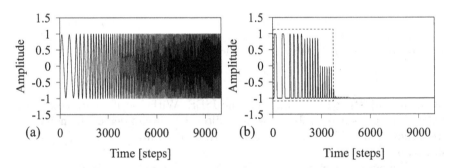

Fig. 5.6. (a) The varying frequencies of an input signal from 100 Hz to 1000 Hz. (b) The output signal of the network. The *dashed frame* is the frequency range (from 100 Hz to approximately 400 Hz) in which the network can detect the auditory signal. Here, the amplitude of the signal which is smaller than a threshold value, e.g., 0.5, is neglected

Figure. 5.6b shows that the network can detect the signal at the frequency up to approximately 400 Hz (see dashed frame) because the amplitude of higher frequency output (>400 Hz) is smaller than a threshold value, e.g., 0.5. To analyze and observe the network behavior, the input signal from the Data Reader is sent to the network implemented on the ISEE and the signals from all neurons are monitored (Fig. 5.7).

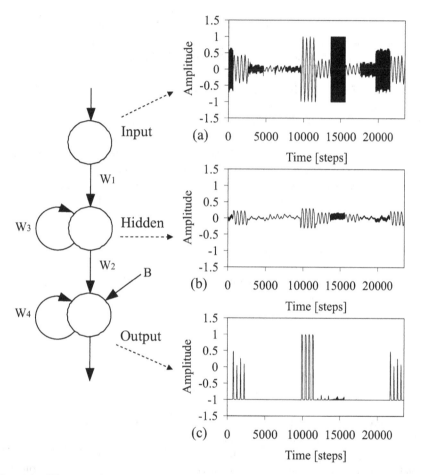

Fig. 5.7. The mixed signals between low- and high-frequency sounds with varying amplitude from all neurons. (**a**) The signals from an input neuron. (**b**) The signals from a hidden neuron. (**c**) The signals from an output neuron

As shown in Fig. 5.7, the first synapse W_1 and the excitatory self-connection W_3 of the hidden unit reduce the amplitude of the input. As a result, the amplitude of a high-frequency sound becomes smaller than the amplitude of a low-frequency sound due to the critical self-connection ($W_3 = 1$) performing as an effective integrator. Afterwards, the signals are again amplified by W_2. Then the bias term B together with the excitatory self-connection W_4 of the output unit shifts the high-frequency sound to oscillate around -0.998 with very small amplitude. Consequently, the network suppresses the high-frequency sound and only the low-frequency sound with a high enough amplitude can pass through the network.

The final step is to implement the advanced auditory network into the mobile system of the walking machine(s) [129]. That is, the auditory signal[2] detected via either the stereo auditory sensor or the auditory–tactile sensor is digitized via the MBoard at a sampling rate of up to 5.7 kHz, and the signal processing network will be programmed on a PDA with an update frequency of $\approx$ 2 kHz. For that, the parameters (weights and a bias) of the advanced auditory network have to be recalculated. An evolutionary algorithm ENS[3] (described in Sect. 3.3), is applied to optimize the parameters of this network. It is implemented on the ISEE, and it receives the input signal for evolutionary process from the Data Reader (cf. Sect. 3.3). The first population consists of the fixed network shown in Fig. 5.8a, and the evolutionary process runs until a reasonable solution is reached, which is determined by the fitness value. The fitness function F that minimizes the mean squared error between the target and the output signals is given by:

$$F = \frac{10}{1 + E} \cdot \tag{5.1}$$

In an ideal situation, the maximum value of F should be 10 while the mean squared error E should be equal to 0. The mean squared error E is evaluated by:

$$E = \frac{1}{N} \sum_{t=1}^{N} (target(t) - output(t))^2 , \tag{5.2}$$

where N is the maximum number of time steps. Here, it is set to $N = 6000$. The target signal is activated by oscillating between around -1 and $+1$ only if a low-frequency signal from 100 to 400 Hz[3] is presented, and it is around -1 in all other cases. This is exemplified in Figs. 5.8b and 5.8c.

After 55 generations, the resulting network had a fitness value of $F = 8.76$, which is sufficient to recognize the low-frequency signal in a desirable frequency range. This is shown in Fig. 5.9.

This evolved advanced auditory network has a similar property as the sensory hair of the spider meaning that both of them act as low-pass filters at the same frequency range. In addition, this preprocessing network can filter the

---

[2] In this set-up, the stereo auditory sensor is used to detect the auditory signal which will be provided to the evolutionary process.

[3] The frequency range is proportional to the frequency range in which the sensory hair of the spider can sense the signal of its prey.

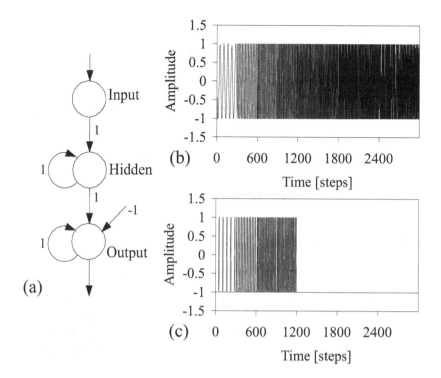

Fig. 5.8. (a) An initial network structure with given weights and a bias. (b) The varying frequencies of an input signal from 100 Hz to 1000 Hz. The input signal is recorded from the physical stereo auditory sensor and then digitized through the ADC channel of the MBoard at a sampling rate of up to 5.7 kHz. (c) A corresponding target signal

noise at high frequencies (>400 Hz) which might occur from the motors of the machine(s) during walking, standing or from the surrounding environments (see demonstration in Chap. 6).

The Sound–Direction Detection Network

In the previous section, the neural preprocessor whose function is similar to a low-pass filter was explained. It is applied to filter undesirable signals coming from, e.g., motors, motions and environments, while it will pass through the sinusoidal sound at a frequency of 200 Hz[4] to trigger a sound tropism.

To discern the direction of the auditory signals for a sound tropism [129], the mentioned ENS[3]-evolutionary algorithm is again applied to find the ap-

---

[4] The selected frequency depends on the distance between two microphones from which the time delay of two signals occurs.

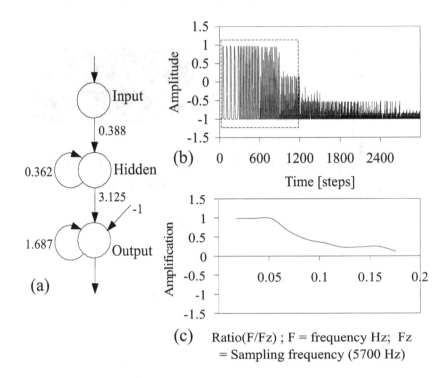

Fig. 5.9. (a) The evolved advanced auditory network applied to the mobile system is optimized by the evolutionary algorithm. It is able to filter the frequency of the auditory signals that are higher than around 400 Hz. (b) The output signal of the network is presented. In the *dashed frame*, there are auditory signals at a low-frequency range approximately between 100 and 400 Hz. (c) The characteristic curve of this network with its cutoff frequency at around 400 Hz where the amplification is smaller than a threshold value, e.g., 0.6

propriate neural network based on the concept of the TDOA [54, 138]. Here, the input signals for the evolved network are detected by the stereo auditory sensor, and they are digitized via the MBoard and then recorded on a PDA at an update frequency of approximately 2 kHz. According to the dimension of the four-legged walking machine AMOS-WD02 and the distance between the fore left and the rear right auditory sensors (see Chap. 4), the maximum time delay between the left and the right signals is equivalent to one-fourth of the wavelength of a frequency of 200 Hz. To evolve the neural network, the same strategy as described above is employed. The initial neural structure is based on the minimal recurrent controller (MRC) [96], and its parameters are shown in Fig. 5.10a. This neural structure consists of two input and two output neurons. The input signals are first filtered via the evolved advanced auditory network; as a result, only noise-free signals at low frequencies can pass through the evolved network. The input signals together with the delay of each are shown in Fig. 5.10b. The fitness function F is determined by (5.1), and the mean squared error E is estimated by:

$$E = \frac{1}{N} \sum_{t=1}^{N} \left[\sum_{i=1}^{2} (target_i(t) - output_i(t))^2 \right].$$

(5.3)

N is equal to 7000, referring to the maximum number of time steps and $i = 1; 2$ refers to the signals on the right and the left, respectively. The target signals are prepared in such a way that they refer to the recognition of a leading signal or to only one active signal. For instance (Fig. 5.10c), Target1 (solid line) is set to $+1$ if the signal of Input1 (I_1) leads the signal of Input2 (I_2) or only I_1 is active indicating that "the sound source is on the right side" and it is set to -1 in all other cases. Correspondingly, Target2 (dashed line) is set to $+1$ in the reverse cases indicating that "the sound source is on the left side".

The network resulting from the evolution after 260 generations has a fitness value of $F = 6.96$, which is sufficient to solve this problem. This sound–direction detection network as well as the inputs and the outputs are presented in Fig. 5.11.

The main feature of this network is its ability to distinguish the direction of incoming signals by observing a leading signal or solely an active signal, and it is easy to implement on the mobile processor because of its uncomplicated neural structure. In addition, its outputs can be directly connected to the neural control module such that it can determine the walking direction of the machine(s); e.g., the machine(s) turns left when the sound source is on the left side and vice versa.

The output neurons of this small network are excited by straight and cross-connections coming from each of the input neurons. There are also ex-citatory self-connections at both output neurons providing hysteresis effects. They allow the switching between two fixed point attractors corresponding to stationary output values of the output neurons, one low and the other high (Fig. 5.12). The strength of a self-connection $W > +1$ determines the width of the hysteresis interval in the input space (see also Sect. 3.2) [96].

However, if the strength of W is too large (for instance, the weight at Output1 $W_1 > 2.0$ and at Output2 $W_2 > 3.5$), then the inputs will not sweep back and forth across the hysteresis domains, with the result that the output signal will oscillate around the high output value when the input signal is activated. This phenomenon is demonstrated in Fig. 5.12, where Output2 versus Input2 for smaller self-connection weights ($W_1 = 2.0$, $W_2 = 3.5$) and larger self-connection weights ($W_1 = 2.206$, $W_2 = 3.872$) are plotted.

Figure 5.12a shows the switching of the output of Output2 (O_2) be-tween almost saturation values (corresponding to the fixed point attractors) while I_2 varies over the whole input interval and I_1 is provided with a de-

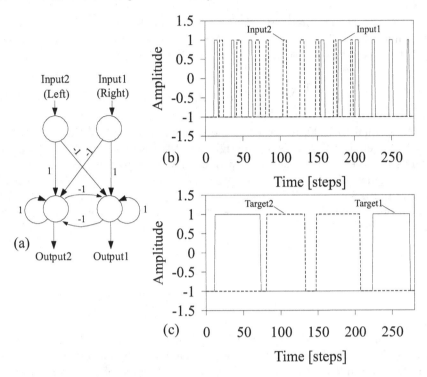

Fig. 5.10. (a) An initial network structure with given weights. (b) The input signals at a frequency of 200 Hz from the right (*solid line*) and the left (*dashed line*) sensors involving the delay between them. At the first period, the sound source is on the right of the walking machine until around 75 time steps it changes to the left. There, only the left sensor can detect the sound, implying that the sound source is a little far away from the right sensor. Then, after around 150 time steps, the walking machine gets closer to the sound source with the result that the right sensor also detects the sound. After around 210 time steps, the sound source is again changed to its right and a little farther away from the left sensor. (c) Target1 (*solid line*) and Target2 (*dashed line*) correspond to the directions of the signals on the right and the left, respectively

lay (Fig. 5.13a). On the other hand, O_2 in Fig. 5.12b jumps and then stays oscillating with very small amplitude around the high output value.

Moreover, one can also see this effect in Fig. 5.13. The output signals corresponding to the different strengths of the self-couplings are plotted for $W_1 = 2.0$, $W_2 = 3.5$, and for the original weights, i.e., $W_1 = 2.206$, $W_2 = 3.872$ (cf. Fig. 5.11a). The sound source is on the left side causing I_1 to follow I_2 with a delay (Fig. 5.13a). Also, the output of Output1 (O_1) is suppressed while O_2 is activated (Figs. 5.13b and 5.13c).

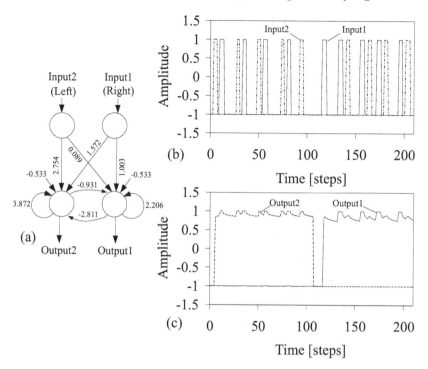

Fig. 5.11. (a) The resulting network called "sound–direction detection network". (b) The input signals from both sensors with the delay between each other. At the first period, the sound source is on the left of the walking machine while it changes to the right after around 110 time steps. (c) During the first period, the signal of Output2 is active and the signal of Output1 is inactive, while the signal of Output2 becomes inactive after around 105 time steps and the signal of Output1 becomes active after around 110 time steps

For the smaller self-connection weights, O_2 oscillates between the low value (approximately -1) and the high value (approximately $+1$) as shown in Fig. 5.13b. For the larger self-connection weights, O_2 oscillates finally with a very small amplitude around the high value above a threshold which can be set from the experiment and which depends on the system, e.g., 0.5 (Fig. 5.13c). Furthermore, one can see that the output neurons form a so-called even loop [150]; i.e., they are recurrently connected by inhibitory synapses (Fig. 5.11a). This configuration guarantees that only one output at a time can be positive, i.e., it functions as a switch, sending the output to a negative value for the delayed input signal. The output signals of this phenomenon can be observed in Fig. 5.11c. By utilizing the phenomena of the larger self-connection weights and the even two-module, one can easily apply the output signals to control the walking direction of the machine(s) for a sound tropism approach.

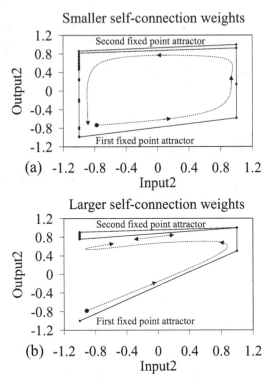

Fig. 5.12. Comparing outputs for different self-connection weights at Output1 and Output2 while I_2 sweeps over the input interval (between -1 and $+1$) and I_1 is given by following Input2 with a delay. (**a**) Varying Output2 for smaller self-connection weights ($W_1 = 2.0$, $W_2 = 3.5$), and (**b**) for larger self-connection weights ($W_1 = 2.206$, $W_2 = 3.872$). *Black spots* indicate the initial output values, which are then following the indicated paths (*dotted line*). There is no hysteresis loop in (**b**) like there is in (**a**); instead it oscillates around the high output value

The Auditory Signal Processing Network

Here, the integration of the evolved advanced auditory network and the sound–direction detection network leads to the conclusive auditory signal processing network [129] (Fig. 5.14). This network has the ability to filter the auditory signals and to discern the direction of the input signals. First, the evolved advanced auditory network filters the sensory inputs (Auditory Input1 and Auditory Input2 in Fig. 5.14) so that only low-frequency sounds can pass through. Second, the outputs from the evolved advanced auditory network are connected to the inputs of the sound–direction detection network. The sound–direction detection network then indicates the direction of the corresponding signals. Subsequently, the output neurons of the sound–direction detection network will be connected to the modular neural controller (described

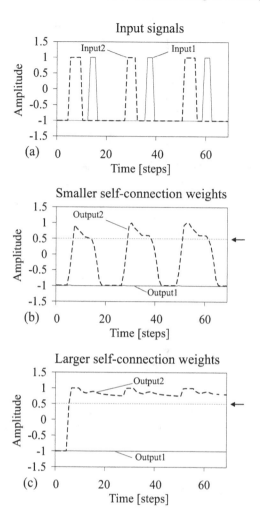

Fig. 5.13. (a) The input signals with a delay; (b) the corresponding oscillating O_2 (*dashed line*) for the smaller self-connection weights ($W_1 = 2.0$, $W_2 = 3.5$) while O_1 (*solid line*) is suppressed. (c) O_2 jumps and stays higher than a threshold (here 0.5 (*arrows*)) for larger self-connection weights ($W_1 = 2.206$, $W_2 = 3.872$)

in Sect. 5.2.3) to make the walking machine(s) turn into the appropriate direction. Eventually, the walking machine(s) will approach and stop near to the source by checking a threshold of the amplitude of the auditory signals (demonstrated in Chap. 6).

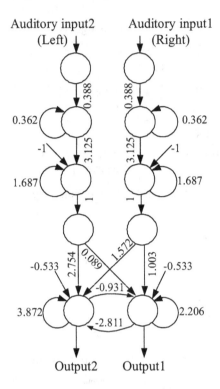

Fig. 5.14. The auditory signal processing network which functions as a low-pass filter circuit and which has an ability to detect the directionality of the corresponding signals. The network is developed to operate at an update frequency of approximately 2 kHz

5.1.2 Preprocessing of a Tactile Signal

Employing the auditory–tactile sensor for sensing an environment in robotic applications will enable mobile robots to detect sound, e.g., at low-frequency (100 Hz), and to avoid collision. These sensor signals consist of an auditory signal and a tactile signal. Both signals are digitized through the line-in port of a sound card at a sampling rate of 48 kHz on a 1-GHz PC, and the preprocessing network will be updated at the same frequency of 48 kHz. The auditory signal is produced via a loudspeaker. It is filtered and recognized by applying the principal advanced auditory network shown in Fig. 5.5. For the tactile signal, it is simulated by sweeping the sensor back and forth over an object. The recorded signal together with its Fast Fourier Transform (FFT) spectrum[5] is exemplified in Fig. 5.15.

---

[5] The FFT spectrum is analyzed by FFTSCOPRE 1.2 software of Physics Dept. at Rutgers University.

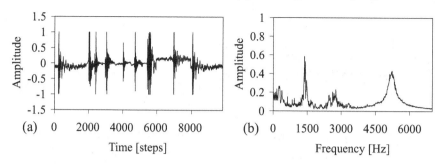

Fig. 5.15. (a) The oscillating peaks are the tactile signal coming from the auditory–tactile sensor. (b) The FFT spectrum displays the compound frequency of the signal. By observing the compound frequency, the first and the second resonance frequencies appear at around 1400 Hz and 5200 Hz

To process the tactile signal, the input signals consisting of the simulated tactile signal and the low-frequency sound at 100 Hz with varying amplitude are prepared on the Data Reader (Fig. 5.16a), and the ENS³ algorithm is applied to evolve for an appropriate neural network via the ISEE. At the beginning only one input and one output unit without connections are given. The ENS³ algorithm then increases or decreases the number of synapses and the hidden units throughout the evolutionary process, which optimizes the parameters at the same time, until the output signals are good enough for a reasonable solution. The fitness function F is chosen in such a way that the evolution minimizes the square error between the target and the output signals; i.e., it is defined by:

$$F = \frac{1}{N} \sum_{t=1}^{N} \left(1 - (target(t) - output(t))^2\right), \tag{5.4}$$

where N is the maximum number of time steps, usually set to $N = 25{,}000$. For an ideal case, the maximum value of F should be $+1$ while the square error between the target and the output signals should be equal to 0. The target signal gives $+1$ if a tactile signal is presented, and -1 in all other cases. This is exemplified in Fig. 5.16b.

The resulting network, a tactile signal processing network, at 800 generations has a fitness value of $F = 0.6$, which is sufficient to recognize the tactile signal (see the recognized output signal in Fig. 5.18d). The network consists of 2 hidden units and 7 synapses as shown in Fig. 5.17. To understand the network behavior, the signals from all neurons are monitored by means of the ISEE, and they are presented in Fig. 5.18.

From observing the signals at the hidden and the output units, the amplitude of low-frequency sound (100 Hz) is reduced at the first hidden unit because of the feedback from the second hidden unit. It becomes smaller than

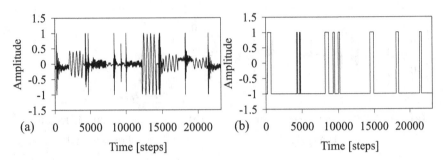

Fig. 5.16. (a) A real input signal coming from the physical sensor. It is mixed between the tactile signal and the low-frequency sound at 100 Hz. (b) The corresponding target function

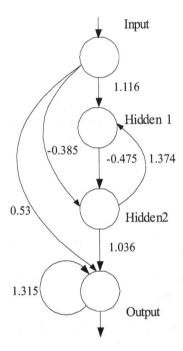

Fig. 5.17. The tactile signal processing network which filters the low-frequency sound. Its output signal follows the tactile signal, which consists of many frequencies and which has hight resonance frequencies as shown in Fig. 5.15b

the amplitude of the tactile signal. Afterwards the amplitudes of both signals are again added in the second hidden unit. Then the excitatory synapse from the input unit together with the excitatory self-connection of the output unit shifts the signal of low-frequency sound to oscillate around −0.78 with small

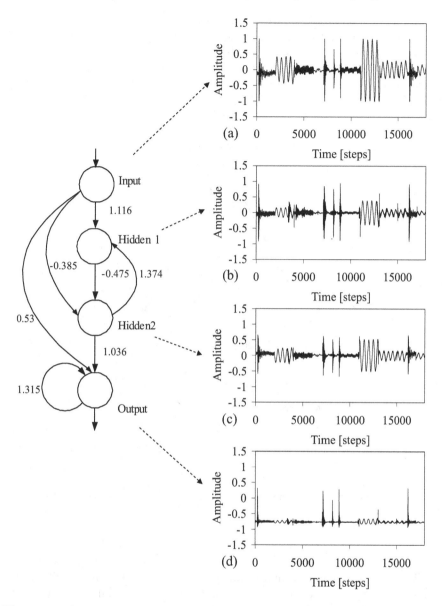

Fig. 5.18. The mixed signals between low-frequency sound (100 Hz) and the tactile signal from all neurons. (**a**) The signals from an input neuron. (**b**), (**c**) The signals from hidden neurons. (**d**) The signals from an output neuron

amplitude. As a result the tactile signal processing network suppresses the signal of low-frequency sound and only the tactile signal is activated.

Here, the combination between the principal advanced auditory network and the tactile signal processing network leads to an auditory–tactile signal processing network. The network is able to distinguish between low-frequency sound and a tactile signal coming from the physical auditory–tactile sensor. This network, consisting of one input unit, three hidden units and two output units, is shown in Fig. 5.19.

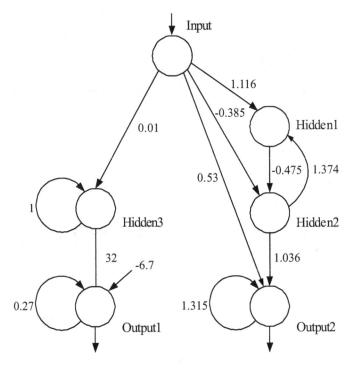

Fig. 5.19. The auditory–tactile signal processing network recognizes low-frequency sound up to 400 Hz (O_1), and the described tactile signal (O_2). It is developed to operate at an update frequency of 48 kHz

The sensor signal is simultaneously provided for the input unit of the principal advanced auditory network and the tactile signal processing network. The signal of output1 (O_1) is active and oscillates between values of approximately 0.998 and −0.997 if low-frequency sound is recognized. And the signal of output2 (O_2) is active if a tactile signal is recognized. Otherwise both output signals are inactive.

5.1.3 Preprocessing of Antenna-like Sensor Data

To obtain an obstacle avoidance behavior by using the sensory information of IR-based antenna sensors, which is digitized via the ADC channels of the MBoard at the sampling rate of up to 5.7 kHz, the preprocessing of the sensor data is required. Here, the property of the MRC [96] is again applied. The MRC has been developed to control a miniature Khepera robot [135], which is a two-wheel platform. The desired preprocessing network was developed and analyzed by using the YARS connecting to the ISEE (cf. Sects. 3.3 and 4.2). The simulation was implemented on a 1-GHz PC with an update frequency of 75 Hz. Eventually, the effective preprocessing network will be transferred to the mobile processor on a physical walking machine.

On the basis of its well-understood functionality [96], the parameters were manually readjusted with the help of the simulation for using it in this approach. First, the weights $W_{1,2}$ from the input to the output units were set to a high value to amplify the sensory signals, i.e., $W_{1,2} = 7$. As a result, under some conditions the sensory noise was eliminated. In fact, these high multiplicative weights drive the output signals to switch between two saturation domains, one low (≈ -1) and the other high ($\approx +1$). Then the self-connection weights of the output neurons were manually adjusted to derive a reasonable hysteresis interval on the input space. The width of the hysteresis is proportional to the strength of the self-connections. This effect determines the turning angle in front of the obstacles for avoiding them, i.e., the wider the hysteresis, the larger the turning angle. Both self-connections are set to 5.4 to obtain the suitable turning angle of the AMOS-WD02. Finally, the recurrent connections between output neurons were symmetrized and manually adjusted to the value -3.55. This guarantees the optimal functionality for avoiding obstacles and escaping from sharp corners. The resulting network is shown in Fig. 5.20.

Generally, two IR-based antenna sensors installed on the forehead of a walking machine (see also Sect. 4.1.3) together with the neural preprocessing above are sufficient to sense the obstacles on the left front and the right front. However, to enhance the avoiding capacity, e.g., protecting the legs of a walking machine from hitting obstacles, like chair or desk legs, one can easily install more sensors at the legs (cf. Sect. 4.1.3), and send all their signals to the corresponding input neurons of the network. For instance, by implementing six sensors on the six-legged walking machine AMOS-WD06, the three sensory signals on each side are simply connected with hidden neurons which are directly connected to the two original output neurons with large weights. To stay in the linear domain of the sigmoid transfer function of the hidden neuron, each sensory signal is multiplied with a small weight, here 0.15, and the bias term (B) is set to determine a threshold value of the sum of the input signals, e.g., 0.2. When the measured value is greater than the threshold in any of the three sensory signals, excitation of the hidden neuron on the corresponding side occurs. Consequently, the activation output of each hidden neuron can

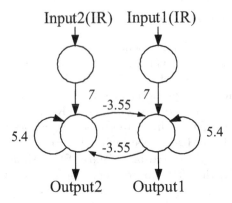

Fig. 5.20. The signal processing network of antenna-like sensors with appropriate weights. The network is developed to operate at an update frequency of 75 Hz. Here, it is applied for controlling the AMOS-WD02

vary in the range between ≈ -0.245 ("no obstacle is detected") and ≈ 0.572 ("all three sensors on the appropriate side simultaneously detect obstacles"). Furthermore, the weights from the hidden to the output units are set to a high value, e.g., 25, to amplify these signals. Again the other parameters (self-connection and recurrent-connection weights of the output neurons) were manually optimized in the similar way as described above. As a result, they are set to 4 and -2.5, respectively. The optimization was first simulated and then finally tested on the AMOS-WD06. The improved structure of this neural preprocessing together with its optimized weights is shown in Fig. 5.21.

In both cases (Figs. 5.20 and 5.21), all sensory signals are linearly mapped onto the interval $[-1, +1]$ before feeding them into the networks, with -1 representing "no obstacles", and $+1$ "a near obstacle is detected". The output neurons of the networks have so-called *super-critical* self-connections (>1.0) which produce a hysteresis effect for both output signals. A strong excitatory self-connection will hold the slightly constant output signal longer than a smaller one, resulting in a larger turning angle to avoid obstacles or corners. To visualize this phenomenon, the network shown in Fig. 5.20 is exemplified and the hysteresis effect is plotted in Fig. 5.22. There, the different weights of excitatory self-connection can be also compared.

In addition, there is a third hysteresis phenomenon involved which is associated to a so-called *even loop* of the inhibitory connection [150] between the two output neurons. Under general conditions, only one neuron at a time is able to get a positive output, while the other one has a negative output, and vice versa. The network shown in Fig. 5.20 is again used to illustrate this phenomenon (Fig. 5.23).

By applying the described phenomena, the sensory noise is eliminated (Fig. 5.22) and the walking machines are able to avoid the obstacles and

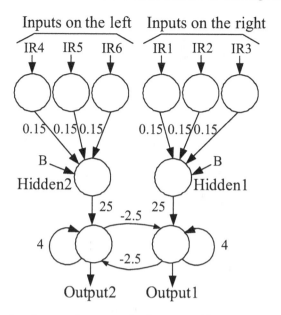

Fig. 5.21. The signal processing network of antenna-like sensors for the six sensory inputs. The network is also developed to operate at an update frequency of 75 Hz. Here, it is applied for controlling the AMOS-WD06

even to escape from a corner and a deadlock situation. The machines will be driven to turn away from the objects with the angle that is determined by the excitatory self-connections of the output neurons. Also, due to the inhibitory synapses, they will determine the direction to which the walking machines should turn when obstacles are detected.

5.2 Neural Control of Walking Machines

To generate the locomotion of walking machines and to change the appropriate motions, e.g., turning left, right or walking backward with respect to sensor signals, an artificial neural network together with the principle of dynamic properties of recurrent neural networks described in Chap. 3 is employed. The neural control [128] for this approach consists of two subordinate networks. One is a neural oscillator network, which generates the rhythmic leg movements, while the other is the velocity regulating network (VRN), which expands the steering capabilities of the walking machines.

5.2.1 The Neural Oscillator Network

Neural oscillators for the walking machines have often been studied [29, 65, 131, 132, 140, 197]. Inter alia, H. Kimura et al. [111] constructed a neural

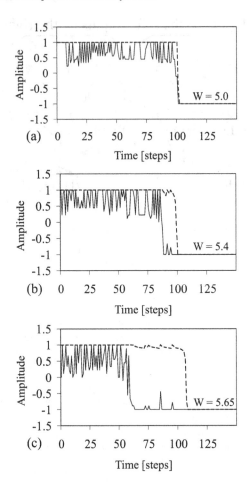

Fig. 5.22. Comparison of the "hysteresis effects" with different self-connection weights at the output neuron. (**a**) The output signal (*dashed line*) decreases from $\approx +1$ to ≈ -1 when the input signal (*solid line*) is inactive (≈ -1). This effect corresponds to a very small turning angle of the walking machine in avoiding an obstacle. (**b**) The output signal (*dashed line*) stays longer at $\approx +1$ and then decreases to ≈ -1 when the input signal (*solid line*) is inactive. This effect corresponds to an appropriate turning angle of the walking machine in avoiding an obstacle. (**c**) The output signal (*dashed line*) stays longest at $\approx +1$ and then decreases to ≈ -1. This effect corresponds to a larger turning angle of the walking machine in avoiding an obstacle

oscillator network with four neurons. The network has been applied to control the four-legged walking machine $TEKKEN$ where each hip joint of the machine is driven by one of the neurons. J. Ayers et al. [12] used a neu-

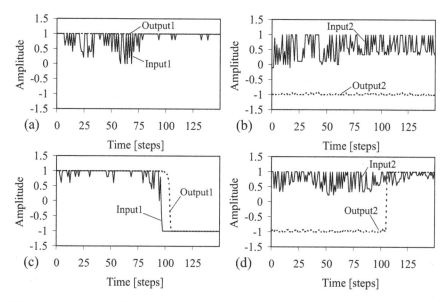

Fig. 5.23. (a)–(d) The input signals (*solid line*) of the sensors and the output signals (*dashed line*) of the output neurons. Due to the inhibitory synapses and the high activity of Output1 (**a**), the Output2 (**b**) is still inactive although Input2 is active. (**c**) and (**d**) show the switching condition between Output1 and Output2 when the activity of Input1 is low, meaning "no obstacles detected" and the activity of Input2 is still high, meaning "obstacles detected". This phenomenon is responsible for escaping from sharp corners as well as deadlock situations

ral oscillator consisting of so-called elevator and depressor synergies. They are arranged as an endogenous pacemaker network with reciprocal inhibition, and are used to generate walking patterns for the eight-legged *Lobster* robot. Here a so-called two-neuron network [154] is employed. It is used as a CPG [101, 113, 171, 198] which follows one principle of locomotion control in walking animals (cf. Sect. 2.3). It generates the rhythmic movement for basic locomotion of the walking machines without the requirement of sensory feedback. The network structure is shown in Fig. 5.24.

The network parameters are experimentally adjusted via the ISEE to acquire the optimal oscillating output signals for generating locomotion of the walking machines. The parameter set is selected with respect to the dynamics of the two-neuron system staying near the Neimark–Sacker bifurcation, where the quasi-periodic attractors occur [154]. Examples of different oscillating output signals generated by different weights and bias terms are presented in Fig. 5.25.

Figure 5.25 shows that such a network has the capability to generate various oscillating outputs depending on the weights and the bias terms. For instance, if the bias terms are small (cf. Fig. 5.25a), the initial output sig-

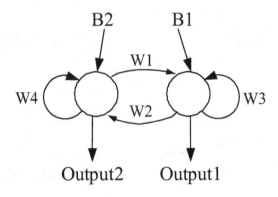

Fig. 5.24. The structure of the two-neuron network

nals will oscillate with a very small amplitude and then the amplitude will
increase during a transient time, while the amplitude of the output signals
for large bias terms is high right from the beginning (cf. Fig. 5.25b). Furthermore, different bias terms also affect the waveform of the output signals.
Different self-connection weights result in different amplitude and waveforms
of the oscillating output signals (compare Figs. 5.25c and 5.25d). To adjust
the oscillating frequency of the outputs, one can also control the connection
weights between two output neurons; i.e., for small connection weights (absolute values), the output signals oscillate at low frequency, while the large
connection weights (absolute values) make the outputs oscillate at high frequency with different waveforms (compare Figs. 5.25e and 5.25f). However,
one can utilize this modifiable oscillating output behavior with respect to the
weights and the bias terms in the field of neural control, e.g., for controlling
the type of walking and the walking speed of legged robots.

Here, the actual parameter set for the network controller is given by $B_1 =
B_2 = 0.01$, $W_1 = -0.4$, $W_2 = 0.4$ and $W_3 = W_4 = 1.5$, where the sinusoidal
outputs correspond to a quasi-periodic attractor (Fig. 5.26). They are used
to drive the motor neurons directly to generate the appropriate locomotion of
the walking machines [74, 128, 130].

The output of neuron 1 (*Output1*) is used to drive all thoracic joints and
an additional backbone joint, and the output of neuron 2 (*Output2*) is used
to drive all basal joints (and all distal joints for a three DOF leg). This oscillator network is implemented on a PDA with an update frequency of 25.6 Hz.
It generates a sinusoidal output with a frequency of approximately 0.8 Hz
(Fig. 5.27) analyzed by the free scientific software package Scilab-3.0.[6]

By using asymmetric connections from the oscillator outputs to corresponding motor neurons, a typical trot gait for a four-legged walking machine
and a typical tripod gait for a six-legged walking machine are obtained which

---

[6] See also: http://scilabsoft.inria.fr/. Cited 18 December 2005.

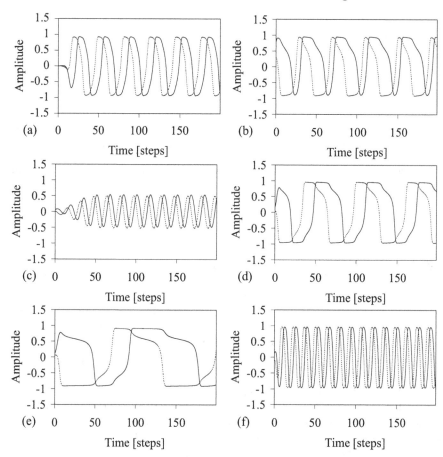

Fig. 5.25. The oscillating output signals of neurons 1 (*dashed line*) and 2 (*solid line*) from the network having different weights and bias terms. (**a**) For small bias terms ($B_1 = B_2 = 0.0001$) while $W_1 = -0.4$, $W_2 = 0.4$ and $W_3 = W_4 = 1.5$. (**b**) For larger bias terms ($B_1 = B_2 = 0.1$) and all weights as in (**a**). (**c**) For smaller self-connection weights ($W_3 = W_4 = 1$) while $W_1 = -0.4$, $W_2 = 0.4$ and bias terms $= 0.01$. (**d**) For larger self-connection weights ($W_3 = W_4 = 1.7$) and all weights together with bias terms as in (**c**). (**e**) For smaller absolute values of connection weights between two output neurons ($W_1 = -0.25$, $W_2 = 0.25$) while $W_3 = W_4 = 1.5$ and the bias terms $= 0.01$. (**f**) For larger absolute values of connection weights between two output neurons ($W_1 = -0.8$, $W_2 = 0.8$) and all weights together with bias terms as in (**e**)

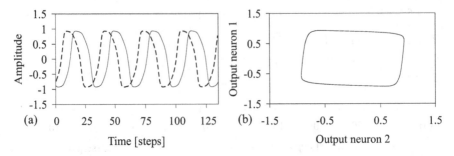

Fig. 5.26. (a) The output signals of neurons 1 (*dashed line*) and 2 (*solid line*) from the neural oscillator network. (b) The phase space with quasi-periodic attractor of the oscillator network which is used to drive the legs of the machines

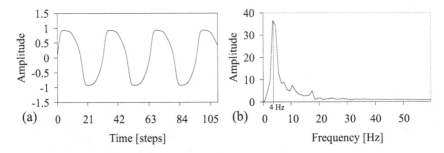

Fig. 5.27. (a) The sinusoidal output generated by the neural oscillator network is recorded for 5 seconds. (b) The FFT spectrum of the recorded sinusoidal output shows that the output has the eigenfrequency around 4 Hz. Then, the walking frequency of the machines can be approximately (4/5) 0.8 Hz

are similar to the gaits of a cat and a cockroach, respectively (described in Sect. 2.3). In a trot gait as well as a tripod gait, (Figs. 5.28 and 5.29), the diagonal legs are paired and move together (see also Sect. 2.3). These typical gaits will enable efficient forward motions.

5.2.2 The Velocity Regulating Network

To change the walking modes, e.g., from walking forwards to walking backwards and from turning left to turning right, the efficient way is to perform a 180-degree phase shift of the sinusoidal signals which drive the thoracic joints. To do so, the VRN is introduced. The network used is modified from [73]. It approximates the multiplication function on two input values $x, y \in [-1, 1]$ (Fig. 5.30). One can optimize this approximation, for instance, by using backpropagation, but it is good enough for the purpose of controlling the machine. Multiplication by higher-order synapses is not used here for consistency reasons.

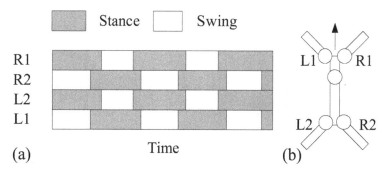

Fig. 5.28. (a) The typical trot gait. The x-axis represents time and the y-axis represents the legs. During the swing phase (*white blocks*) the feet have no ground contact. During the stance phase (*gray blocks*) the feet touch the ground. (b) The orientation of the legs of the AMOS-WD02

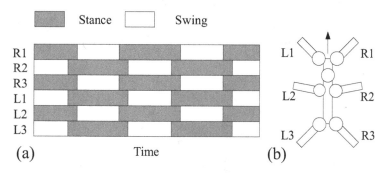

Fig. 5.29. (a) The typical tripod gait. The x-axis represents time and the y-axis represents the legs. During the swing phase (*white blocks*) the feet have no ground contact. During the stance phase (*gray blocks*) the feet touch the ground. (b) The orientation of the legs of the AMOS-WD06

For this purpose the input x is the oscillating signal coming from the neural oscillator network to generate the locomotion and the input y is the sensory signal coming from the neural preprocessing network, e.g., the auditory signal processing, the tactile signal processing or the signal processing of antenna-like sensors, to drive the corresponding behavior. Figure 5.31a represents the network, consisting of four hidden neurons and one output neuron. Figure 5.31b shows that the output signal gets a phase shift of 180 degrees, when the sensory signal (input y) changes from -1 to $+1$ and vice versa.

Because the VRN behaves qualitatively to a multiplication function, it then should also be able to increase and decrease an amplitude of the oscillating signal. To explore the behavior of this network, the fixed oscillating signal is connected to the input x of the network while the input y gets constant

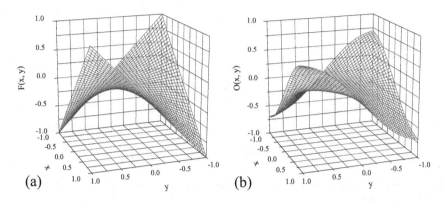

Fig. 5.30. (a) The multiplication function $F(x, y) = x \times y$, and (b) its approximation $O(x, y)$ of the VRN with average mean square error $(e^2) \approx 0.0046748$. The output O of the neuron is given by the sigmoidal transfer function tanh; therefore the suitable input values x, y are in the range of $[-1 \cdots 1]$

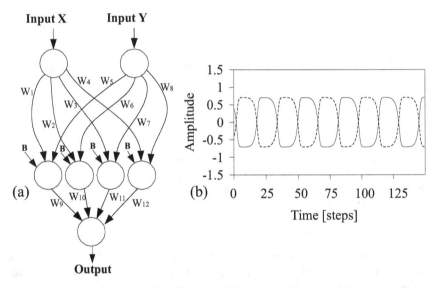

Fig. 5.31. (a) The VRN with four hidden neurons, where the parameter set for the network is given by $W_1 = W_3 = W_5 = W_8 = 1.7246$, $W_2 = W_4 = W_6 = W_7 = -1.7246$, $W_9 = W_{10} = 0.5$, $W_{11} = W_{12} = -0.5$, and the bias terms B are all equal to -2.48285. (b) The output signal (*solid line*) when the input y is equal to $+1$ and the output signal (*dashed line*) when the input y is equal to -1

input values to be multiplied with the oscillating signal. The resulting outputs for the different y-input values which are monitored via the ISEE are shown in Fig. 5.32.

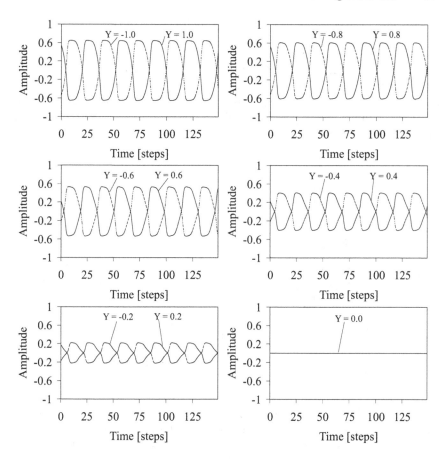

Fig. 5.32. The output signal (*solid line*) when the input y is equal to positive value and the output signal (*dashed line*) when the input y is equal to negative value. The different given values of the input y result in the different amplitudes of the output signal

From Fig. 5.32, it can be seen that the network is not only able to make a 180-degree phase shift of the oscillatory output signal but, using the input y, it can also modulate its amplitude. Especially the amplitude of the output will be 0 if the given input y is equal to 0. This function of the network enables the machines to perform different motions by making a 180-degree phase shift of the oscillatory signal. It even can stop the walking machines by setting the input y to 0. Furthermore, the different amplitudes of the oscillating signal will affect the walking velocity of the machines; i.e., the higher amplitude of the signal the faster they walk and vice versa.

To compare the effect of the different amplitudes of the oscillating signal with the walking velocity of the machines, the VRNs together with the neural oscillator are implemented on the mobile processor of the physical four-legged walking machine AMOS-WD02. The network is updated with 25.6 Hz. To determine the walking velocity of the machine depending on the y-input values, the time needed to cover a fixed distance (1 m) was measured several times for every y-input. The average velocity values for the different y-input are displayed in Fig. 5.33.

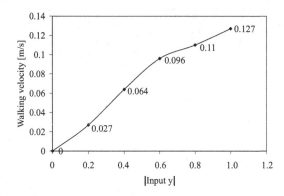

Fig. 5.33. Comparison of the walking velocity with different input y of the VRN

Figure 5.33 shows that the amplitude of the oscillating signal influences the walking velocity of the machine because the higher amplitude provides the larger angle of the thoracic joints in moving forwards and backwards; e.g., $|$input $y| = 0.2$ generates a very small amplitude (Fig. 5.32) of the output resulting in a slow motion (0.027 m/s), on the other hand $|$input $y| = 1.0$ causing a high amplitude and a fast motion (0.127 m/s). Therefore, the VRN together with the neural oscillator can accelerate, decelerate or stop the motion of the walking machines simply driven by sensor input through the so called y-input of the VRN.

5.2.3 The Modular Neural Controller

The integration of two different functional neural modules, *the neural preprocessing and the neural control (the neural oscillator network and the velocity regulating networks)*, gives the effective modular neural controller to generate reactive behaviors. One oscillating output signal from the neural oscillator network is directly connected to all basal joints, while the other is connected to the thoracic joints, only indirectly, passing through all hidden neurons of the VRNs through their x-inputs (Fig. 5.31a). The output signals of the neural preprocessing module go to Input1 (I_1) and Input2 (I_2) of the

VRNs (Figs. 5.34 and 5.35). Thus, the rhythmic leg movements are generated by the neural oscillator network, and the steering capabilities of the walking machines are realized by the VRNs in accordance with the outputs of the neural preprocessing module. The structure of this controller and the location of the corresponding motor neurons on the four-legged walking machine AMOS-WD02 are shown in Fig. 5.34.

The same controller can also be applied to control even more complex systems, e.g., the six-legged walking machine AMOS-WD06 with additional distal joints, *without changing the internal parameters and the structure of the controller* (compare the dashed frame in Figs. 5.34 and 5.35). Only motor and sensory neurons are added. One output of the neural oscillator network drives all basal and distal joints. The other drives all thoracic joints by connecting through all hidden neurons of the VRNs. The network structure and the corresponding positions of the motor neurons of the AMOS-WD06 are shown in Fig. 5.35.

5.3 Behavior Control

Now we shall look at those neural modules which are used on the mobile system,[7] where two signal processing networks have been employed. One is the signal processing network of the antenna-like sensors, and the other is the auditory signal processing network. Utilizing the modular concept, each signal processing network from neural preprocessing module is selected and connected to the neural control module of the four- or six-legged walking machine. Thus different behavior controllers, for instance, obstacle avoidance and sound tropism controllers, can be created by taking this concept into account. In order to achieve more complex behavior in the walking machine(s), a sensor fusion technique is also applied, whereby it has to cooperate or manage the sensory signals.

5.3.1 The Obstacle Avoidance Controller

Our obstacle avoidance controller is constructed from two modules: the signal processing network of antenna-like sensors from the neural preprocessing module and the modular neural controller from the neural control module (Fig. 5.36).

The controller generates the obstacle avoidance behavior where the modular neural controller together with the preprocessing network will enable the machines to walk as well as steer the walking directions of the machines by changing the rhythmic leg movements at the thoracic joints in accordance

---

[7] The controllers are implemented on the PDA with an update frequency of 25.6 Hz, and the sensor signals are digitized via the MBoard at the sampling rate of up to 5.7 kHz.

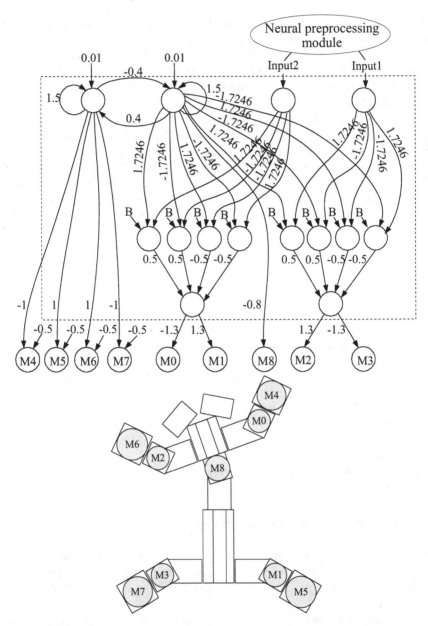

Fig. 5.34. The modular neural controller of the four-legged walking machine AMOS-WD02. It generates a trot gait which is modified when I_1 or I_2 is changed by the sensory signals. The bias terms B of the VRNs are all equal to -2.48285. The outputs from the neural preprocessing module are directly connected to the input neurons (I_1, I_2) of the neural control module (*dashed frame*)

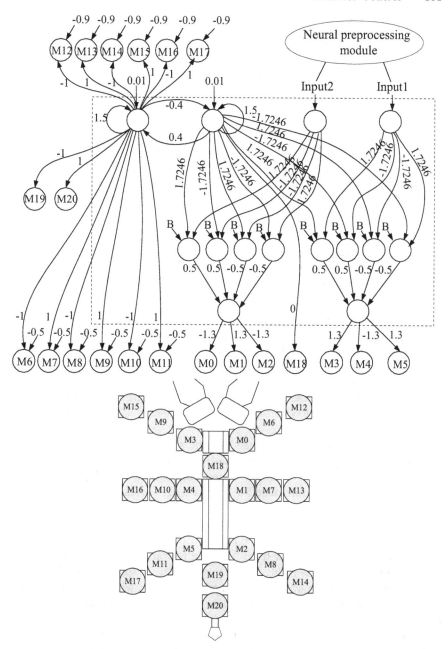

Fig. 5.35. The modular neural controller of the six-legged walking machine AMOS-WD06. The bias terms B of the VRNs are again all equal to -2.48285

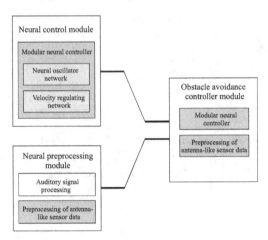

Fig. 5.36. The modular architecture of the obstacle avoidance controller consists of the neural preprocessing and control modules. The preprocessing of antenna-like sensor data from the neural preprocessing module is selected and linked to the modular neural controller (of a four- or six-legged walking machine)

with the sensory signals. Furthermore, the controller even has the capability to prevent the walking machines from getting stuck in a corner or in a deadlock situation because of the hysteresis effects provided by the recurrent structure of the preprocessing network (cf. Figs. 5.22 and 5.23). The structure of the obstacle avoidance controller for the four-legged walking machine [128] is shown in Fig. 5.37.

The same concept can be applied to the six-legged walking machine by connecting the preprocessing of antenna-like sensor data (cf. Fig. 5.21) to the modular neural controller of the six-legged walking machine. The structure of the obstacle avoidance controller for the six-legged walking machine [130] is shown in Fig. 5.38.

As a result, the output signals of the preprocessing network together with the VRNs determine and switch the behavior of the walking machines; for instance, switching the behavior from "walking forward" to "turning left" when there are obstacles on the right, and vice versa. The output signals also determine the direction of the walking machines. Practically, which way they should turn depends on which antenna-like sensor signals have been previously active. In special situations, like walking toward the wall, the antenna-like sensors of the fore left and the fore right might get positive outputs at the same time, and, because of the VRNs, the walking machines are able to walk backward. While walking backward, one of the sensory signals might be still active while the other might be inactive. Correspondingly, the walking machines will turn into the opposite direction of the active signal and they can finally leave the wall.

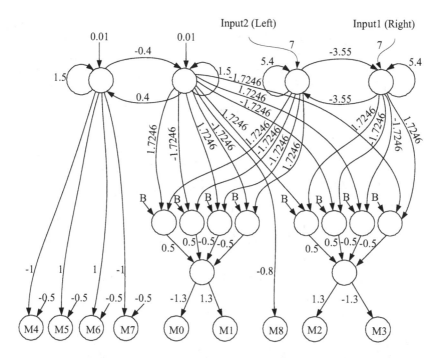

Fig. 5.37. The controller is built from a combination of the preprocessing of antenna-like sensor data and the modular neural controller of the four-legged walking machine. The left and the right signals of the antenna-like sensors are directly connected to input neurons of the signal processing network

5.3.2 The Sound Tropism Controller

The controller that generates a sound tropism inspired by the prey capture behavior of the spider *Cupiennius salei* is built by realizing a modular concept.; i.e., the auditory signal processing of the neural preprocessing module is assembled with the modular neural controller. The modular architecture of the sound tropism is drawn in Fig. 5.39.

In the sound tropism controller [129], the auditory signal processing acts as a low-pass filter by passing through the specific frequency sound (200 Hz) to trigger a behavior and by filtering all high-frequency noise (>400 Hz). Additionally, it can discern the direction of the signals while the modular neural controller has the capacity to enable and to control the motions of the walking machine(s). Consequently, the desired different walking patterns which respond to a switch-on sound source are performed. That is, the machine(s) walks straight, turns toward a switched-on sound source, then makes an approach and then stops near to it by checking the amplitude of the auditory

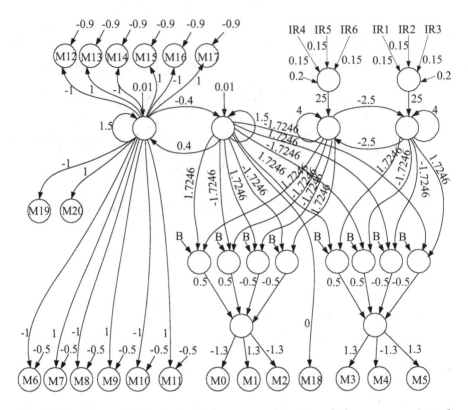

Fig. 5.38. The controller is built from a combination of the preprocessing of antenna-like sensor data and the modular neural controller of the six-legged walking machine

signals. The controller structure of the four-legged walking machine, generating a so-called sound tropism is presented in Fig. 5.40.

However, due to the modular concept, the sound tropism controller can be modified to be implemented on the six-legged walking machine as well. This can be achieved by connecting the auditory signal processing network with the modular neural controller of the six-legged walking machine.

5.3.3 The Behavior Fusion Controller

The combination of the mentioned controllers leads to a versatile artificial perception–action system. This means that the resulting controller can produce different reactive behaviors in accordance with the sensory inputs. For instance, the sensory signals of antenna-like sensors should generate a negative tropism, while the auditory signals should generate a positive tropism so that the machine(s) follows a sound source but avoid obstacles. The mod-

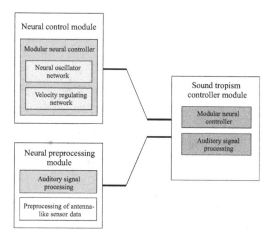

Fig. 5.39. The modular architecture of the sound tropism consists of the neural preprocessing and control modules. The auditory signal processing of the neural preprocessing module is chosen to connect to the modular neural controller (of a four- or six-legged walking machine)

ular architecture of the controller generating different reactive behaviors is illustrated in Fig. 5.41.

As shown in Fig. 5.41, two signal processing networks of different sensory inputs together with the modular neural controller are employed to construct a so-called behavior fusion controller. Both sensory signals have to be managed before directing them to the modular neural controller to execute a behavior. To do so, a fusion technique for the sensor signals is required. It will combine two different sensor data, namely the auditory signals coming from the stereo auditory sensor and IR signals coming from the antenna-like sensors. The preprocessed signals of both sensors go into a fusion procedure in parallel. It manages all input signals and provides only two output signals which are later connected to the modular neural controller. Consequently, the modular neural controller sends the command to the motor neurons of the walking machine(s) to activate the desired behavior. The controller structure is shown in Fig. 5.42.

This fusion procedure consists of two methods: a look-up table and time scheduling. The look-up table method for this approach is used like a table that manages the input signals concerning their priorities. To manage the priority of the sensory signals, the IR signals are desired to have higher priority than the auditory signals. If the obstacles and the auditory signals are detected at the same time, the controller will execute the obstacle avoidance behavior instead of the sound tropism. The sound tropism is performed if and only if the obstacles are not detected. From these statements, 16 actions in accordance with 4 sensory inputs can be executed, where 2 of them come from

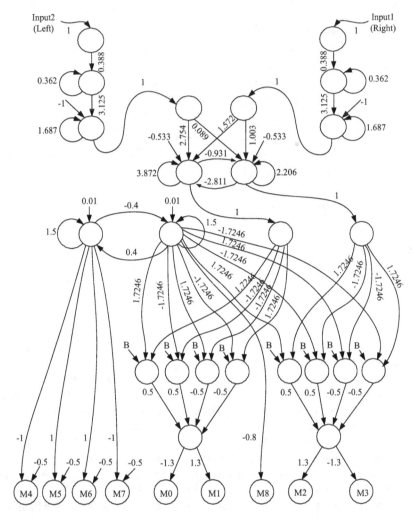

Fig. 5.40. The controller is built from a combination of the auditory signal processing network and the modular neural controller of the four-legged walking machine. The left and the right signals of the auditory sensors are directly connected to input neurons of the auditory signal processing network

the stereo auditory sensor, and the other 2 come from the antenna-like sensors. The driven actions are shown in Table 5.1, where IRR and IRL indicate IR signals of the right and the left antenna-like sensors after preprocessing, respectively; AR and AL indicate auditory signals of the right and the left auditory sensors after preprocessing, respectively; +1 and −1 indicate the active and the inactive signals, respectively.

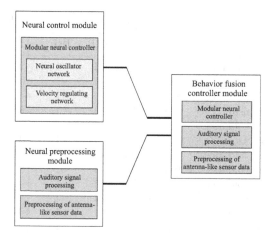

Fig. 5.41. The modular architecture of the behavior fusion controller, which completes the perception–action systems, consists of the neural preprocessing and control modules. The auditory signal processing and the preprocessing of antenna-like sensor data are selected to connect to the modular neural controller (of a four- or six-legged walking machine)

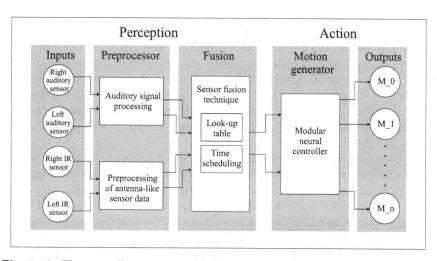

Fig. 5.42. The controller structure of behavior control compounds of preprocessing sensory signals, a sensor-fusion procedure and the motion generator. It filters the input signals at preprocessing channels and then it integrates and manages the signals at the fusion channel. Finally, it sends the output commands to the motor neurons M_n via the motion generator; where $n = 3$ is the number of thoracic motor neurons of the four-legged waking machine and $n = 5$ is the number of thoracic motor neurons of the six-legged walking machine

Table 5.1. The look-up table to manage the sensory inputs

Behavior	Actions	IRR	IRL	AR	AL
Obstacle avoidance	Turn Left	+1	−1	−1	+1
Sound tropism	Turn Left	−1	−1	−1	+1
Obstacle avoidance	Turn Right	−1	+1	−1	+1
Sound tropism	Turn Left	+1	+1	−1	+1
Obstacle avoidance	Turn Left	+1	−1	+1	+1
Sound tropism	Forward	−1	−1	+1	+1
Obstacle avoidance	Turn Right	−1	+1	+1	+1
Obstacle avoidance	Backward	+1	+1	+1	+1
Obstacle avoidance	Turn Left	+1	−1	+1	−1
Sound tropism	Turn Right	−1	−1	+1	−1
Obstacle avoidance	Turn Right	−1	+1	+1	−1
Sound tropism	Turn Right	+1	+1	+1	−1
Obstacle avoidance	Turn Left	+1	−1	−1	−1
Default behavior	Forward	−1	−1	−1	−1
Obstacle avoidance	Turn Right	−1	+1	−1	−1
Obstacle avoidance	Backward	+1	+1	−1	−1

As a result, there are only two situations where the machine(s) is driven to walk forward. One is when the obstacles are not detected (IRR and IRL = −1) and the auditory signals are active (AR and AL = +1) at the same time, which rarely occurs because auditory sensors are installed on the AMOS-WD02 in the diagonal locations. The other one is the normal condition (default behavior) in which the obstacles and auditory signals are not detected. Thus, the machine(s) might have difficulties approaching the sound source although eventually it can reach and stop near the source (see demonstration in Sect. 6.2.2).

To overcome the described problem, a time scheduling technique is added into the fusion procedure. It switches between two behavioral modes, namely obstacle avoidance mode (Om) and composite mode (Cm), made up of the sound tropism and the obstacle avoidance behavior. The obstacle avoidance mode is the mode in which the machine(s) cannot react to the auditory signals although the signals can be detected. On the other hand, the composite mode is the mode in which the walking machine(s) can react to the auditory signals and can also avoid the obstacles, but the performed action is checked by the look-up table method (see Table 5.1).

Two behavioral modes are executed with different time scales and are constantly repeated until the processor time is expired; for instance, the obstacle avoidance mode is primarily executed at approximately 3.2 s of the total (approximately 16.9 s) while the composite mode is suspended. After that the obstacle avoidance mode becomes a suspension and the composite mode becomes executable for approximately 13.7 s. The process will repeatedly run until the processor time is terminated (e.g., ≈ 15 minutes). The different time scales of the behavioral modes can be calibrated and optimized by the experiment depending on each system. Normally, the time scale of a composite mode should be larger than the obstacle avoidance mode. The time scheduling diagram is presented in Fig. 5.43.

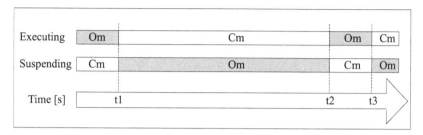

Fig. 5.43. The time scheduling diagram of the sensor fusion technique. At the start, the obstacle avoidance mode (Om) is executed for $t_1 \approx 3.2\,$s while the composite mode (Cm) is suspended. After that the Cm becomes executable for $\approx 13.7\,$s and the Om becomes suspension at the same time. At time t_2 ($\approx 16.9\,$s), the process is complete. It then repeats itself by executing the Om and suspending the Cm. The switching between executing and suspending the Om and Cm is performed until the processor time is terminated, e.g., $\approx 15\,$minutes

From the described strategy, the walking machine(s) walks forward if no obstacles and no sound are detected. It then orients its movement into the direction of the sound source if the sound is detected, with no obstacles, during execution of the composite mode. After that, it will be able to walk forward for a while when the obstacle avoidance mode becomes active and no obstacles are detected. Eventually, it will approach the sound source and stop near it.

To prevent the walking machine(s) from colliding with the sound source while approaching it, the amplitude of the auditory signals must be closely observed and checked. If the amplitude is higher than the threshold, then the input signals ($Input_1$ and $Input_2$) which are connected to the modular neural controller are set to 0. Consequently, the signals of the thoracic motor neurons are inhibited causing the walking machine(s) to stop at a distance determined by the amplitude threshold of the auditory signals. The structure of the behavior fusion controller of the four-legged walking machine together with the specific parameters is given in Fig. 5.44.

To reproduce the sound tropism in the six-legged walking machine, the modular neural controller of the four-legged walking machine can be replaced by the controller of the six-legged walking machine. However, using the behavior fusion controller together with a sensor fusion technique, the output signals from the preprocessing channels are prioritized and coordinated in the fusion channel before sending out the final sensory signals to drive the behavior through the modular neural controller. On the one hand, the output

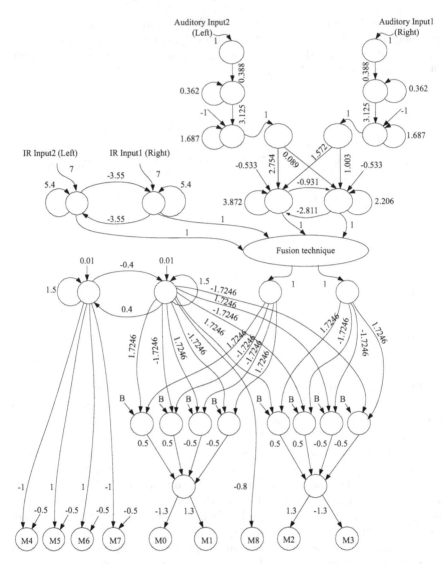

Fig. 5.44. The behavior fusion controller for generating the different reactive behaviors of the four-legged walking machine

signals of the preprocessing of the antenna-like sensors are clarified as the negative response to the stimulus which drives the machine(s) to turn away from the obstacles. On the other hand, the output signals of the auditory signal processor act as the positive response to the stimulus which drives the machine(s) to turn toward the sound source.

5.4 Conclusion

In this chapter, artificial perception–action systems which perform the different reactive behaviors of the walking machine(s) were introduced. They are built from a combination of neural preprocessors for sensor data processing and the neural control for locomotion of the walking machines. The neural preprocessing and control are achieved by applying the dynamic properties of the recurrent neural networks. The optimization of the parameters of the neural preprocessing is achieved by an evolutionary algorithm. Three different types of neural preprocessing modules are presented: auditory signal processing, preprocessing of the antenna-like sensor data and the tactile signal processing. Using the stereo auditory sensor, the sound is processed by the auditory signal processing network acting as a low-pass filter and also discerning the direction of the signals. For the preprocessing of the antenna-like sensor data, it has the capability to eliminate the sensory noise and to control the walking direction of the machines by utilizing the hysteresis effect. Applying the auditory–tactile sensor for collision detection and low-frequency sound detection, the signal coming from the tactile channel is recognized by the tactile signal processing network while the low-frequency sound is recognized by a part of the auditory signal processing network.

In order to obtain the different behavior controllers of the walking machine(s), e.g., an obstacle avoidance controller and a sound tropism controller, each neural preprocessing module of the corresponding sensory signals can be connected to a neural control called a "modular neural controller". This modular neural controller composes the neural oscillator network, which generates the rhythmic leg movements as the CPG, and the VRNs, which expand the steering capabilities of the walking machines. Eventually, the combination of the neural preprocessing and neural control, including the additional sensor fusion technique, will lead to an effective behavior fusion control which enables the walking machine(s) to respond to environmental stimuli, e.g., wandering around, avoiding obstacles and moving toward a sound source.

6

Performance of Artificial Perception–Action Systems

In order to test the capabilities of the artificial perception–action systems, several experiments were carried out. First, the signal processing networks were tested with the simulated signals and the real sensor signals. Afterwards the physical sensors, the neural preprocessing and the neural control were all together implemented on the physical walking machine(s) to demonstrate different reactive behaviors.

6.1 Testing the Neural Preprocessing

This section describes the experiments which show the performance of the neural preprocessing by testing it with the simulated data and the physical sensor data. Afterwards the effective neural preprocessing together with the physical sensor systems, known as an artificial perception part, will be applied for behavior control of the reactive walking machine(s).

6.1.1 The Artificial Auditory–Tactile Sensor Data

An artificial auditory–tactile sensor was built, together with its preprocessing networks. The purpose of this sensor system is to provide environmental information for a sensor-driven system in wheeled robots as well as in walking machines. Here the performance of the auditory signal processing of the sensor, which helps recognizing low-frequency sound (e.g., 100 Hz) as well as eliminating unwanted noise, was previously tested. Afterwards the capability of the tactile signal processing of the sensor, which should detect a real tactile signal, was presented. Thus, using the sensor coupled with the effective signal processing will enable the sensor system to distinguish and to recognize the real auditory and tactile signals.

First of all, the signal processing networks of the auditory–tactile sensor, simple and principal advanced auditory networks, were created on the ISEE (cf. Sect. 3.3) running on a 1-GHz PC at an update frequency of 48 kHz. They

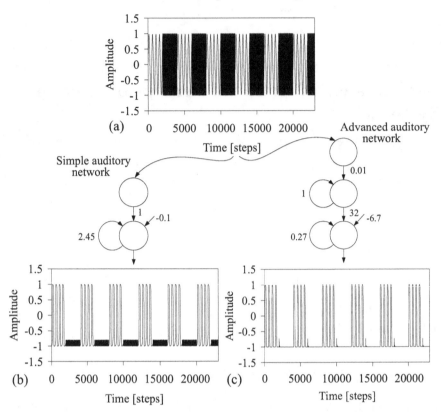

Fig. 6.1. (a) The simulated input consisting of two different frequencies (100 Hz and 1000 Hz). (b) The corresponding output of the simple auditory network. (c) The corresponding output of the principal advanced auditory network. All figures have the same scale in the x-axis and the y-axis

were then tested with a simulated sinusoidal input[1] having constant-amplitude signals and consisting of two different frequencies, one low (100 Hz) and the other high (1000 Hz). Figure 6.1 shows the ideal noise-free input signals and the output signals of the networks.

The same procedure was done with the noisy signals; i.e., the low- and high-frequency sounds were produced by a powered loudspeaker system (30 watts) and recorded via the sensor from a real environment (Fig. 6.2). The output signals of the sensor were digitized through the line-in port of a sound card at a sampling rate of 48 kHz.

---

[1] The signals were simulated with an update frequency of 48 kHz by the wave generator software of Physics Dept. of Rutgers University. They were buffered into the simulator called "Data Reader".

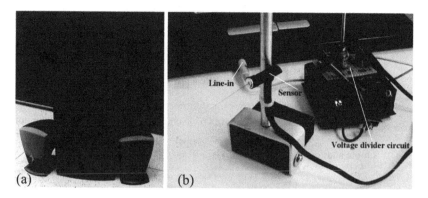

Fig. 6.2. The experimental equipment to record real auditory signals via the auditory–tactile sensor. (**a**) The loudspeaker producing the low- and high-frequency sounds. (**b**) The auditory–tactile sensor system consisting of the sensor, a voltage divider circuit and a PC having a line-in port

The recorded signals with varying amplitudes consisted of the low- and high-frequency sounds, 100 Hz and 1000 Hz, respectively. These signals were filtered through the networks that behaved like a low-pass filter. That is, the simple auditory network can almost pass through the low-frequency sound having the highest amplitude. Although there is some remaining noise from the high-frequency sound which has the highest amplitude input, the noise can be ignored because all of it is low amplitude output (e.g., below −0.50). On the other hand, the advanced auditory network has more capability to pass through some other lower amplitudes of the low-frequency sound. These processes are presented in Fig. 6.3.

Finally, the sensor was applied to a real walking machine, i.e., one sensor was implemented on one leg of the walking machine AMOS-WD02 (Fig. 6.4) and the signals were again recorded through the line-in port.

There were three different settings for recording signals to test the networks. The first setting was that the walking machine was switched on in the initial standing position. The next setting was to let the machine walk, and the last setting was to generate the sound at 100 Hz while the machine was walking (the experimental set-up was similar to the set-up shown in Fig. 6.7). The signals of all settings are shown in Fig. 6.5a and the resulting signals after filtering by the simple and principal advanced auditory networks are presented in Figs. 6.5b and 6.5c, respectively.

Comparing the performance between the simple and principal advanced auditory networks, Fig. 6.1 shows that both networks are able to recognize the low-frequency signal when the signal is noise-free with high-constant amplitude. For the noisy signals shown in Fig. 6.3, the principal advanced auditory network is more robust and it can detect the low-frequency sound with sufficiently high amplitude while the simple auditory network can detect only

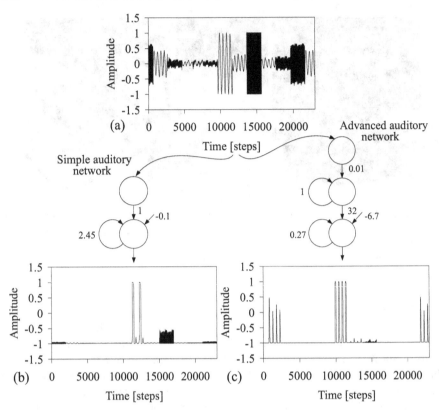

Fig. 6.3. (a) The real input signals (consisting of 100 Hz and 1000 Hz) with varying amplitudes recorded via the physical auditory–tactile sensor. (b) The corresponding output signals of the simple auditory network. (c) The corresponding output signals of the advanced auditory network. All figures have the same scale in the x-axis and the y-axis

the highest amplitude (Figs. 6.3b and 6.3c). Additionally, both networks can filter noise coming from the motor sound of the walking machine in motion as well as in a standing position; but, only the principal advanced auditory network can recognize the low-frequency sound while the machine is walking and simultaneously listening to the sound.

Therefore, the principal advanced auditory network is appropriate for further applications, e.g., one can blend the principal advanced auditory network with the tactile signal processing network to acquire a so-called auditory–tactile signal processing network of the auditory–tactile sensor (described in Sect. 5.1.2). Moreover, one can even combine the evolved advanced auditory network developed on the basis of the principal advanced auditory network with the sound–direction detection network and then implement the combined network, known as "auditory signal processing network", on the mobile sys-

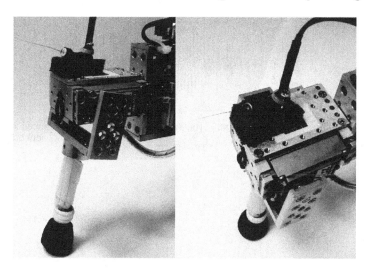

Fig. 6.4. The auditory–tactile sensor was installed on one leg of the AMOS-WD02. The sensor has the extension part (the whisker of a real mouse) around 4.0 cm from the leg

tem of the walking machine(s) to perform the sound tropism as described in the previous chapter.

To show the capability of the auditory–tactile signal processing network in detecting and distinguishing a sound and a tactile signal, the network was again implemented on the ISEE at an update frequency of 48 kHz and received the input data via the Data Reader. The experiment was performed with mixed signals between the low-frequency sound (100 Hz) with varying amplitudes and the tactile signal where the signals come from the physical sensor. The low-frequency sound was generated by a loudspeaker system and the tactile signal was produced in a simple way; that is, the sensor was manually moved back and forth across an object. The input data were recorded through the line-in port and then buffered into the Data Reader connected to the ISEE. The input and output signals of the auditory–tactile signal processing network are exemplified in Fig. 6.6.

As a result, the signal of Output1 (O_1) is shifted to around −0.9 when the low-frequency sound is not presented and the signal of Output2 (O_2) is shifted to around −0.77 when the tactile signal is not presented. Both output signals will be activated in reverse cases. The output signals (O_1, O_2, see Figs. 6.6b and 6.6c) of the network prove that the evolutionary algorithm ENS[3] is able to construct an effective network for a signal processing approach by utilizing discrete-time dynamical properties of recurrent neural networks.

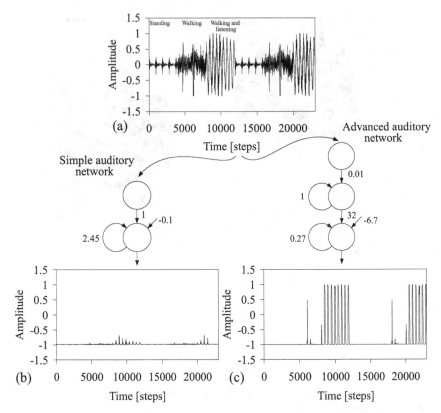

Fig. 6.5. (a) The real input signals consisting of three different conditions (standing, walking and walking while listening to the sound). (b) The corresponding output signal of the simple auditory network. (c) The corresponding output signal of the principal advanced auditory network. All figures have the same scale in the x-axis and the y-axis

6.1.2 The Stereo Auditory Sensor Data

The supplemental application of an evolved advanced auditory network (cf. Fig. 5.9) was attempted. It shall be used for producing a sound tropism in the walking machine(s). The network is developed to work on a mobile system which consists of a PDA having an Intel (R) PXA255 processor coupled with the MBoard. They communicate via an RS232 interface at 57.6 kbits/s.

All of the forthcoming experiments were carried out on the mobile system of the four-legged walking machine AMOS-WD02. All tested signal processing networks were programmed on the PDA, which has an update frequency of up to 2 kHz. The sensor signals coming from the fore-left and the rear-right auditory sensors were digitized via the ADC channels of the MBoard at a sampling rate of up to 5.7 kHz.

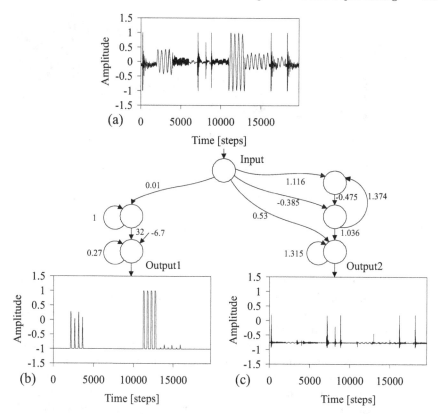

Fig. 6.6. (a) The mixed signals between the low-frequency sound (100 Hz) with varying amplitudes and the tactile signal. They were recorded via the auditory–tactile sensor. (b) The response of the network to the low-frequency sound. (c) The response of the network to the tactile signal. Both outputs are active only for sound and the tactile signal. All figures have the same scale in the x-axis and the y-axis

The first attempt was to test the evolved advanced auditory network with the unexpected noise of three different conditions: standing, walking without listening to sound and walking while listening to sound. In the last condition, the walking machine was initially placed in front of a loudspeaker at a distance of 30 cm, and the low-frequency sound at 200 Hz having a basic sine shape was generated via the loudspeaker (Fig. 6.7). The sound with this frequency was selected for testing because it was later applied to trigger the sound tropism.

The inputs and the corresponding outputs of the network from the different conditions are illustrated in Figs. 6.8 and 6.9. As a result, Figs. 6.8 and 6.9 show that the network is able to remove most unwanted noise, e.g., the motor sound of the walking machine during standing and unpredictable noise during the walk. This is because some of them have a low-amplitude signal and most of them vibrate at high frequencies. However, some unwanted noise still

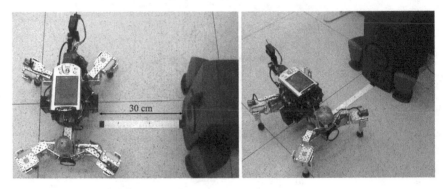

Fig. 6.7. The walking machine AMOS-WD02 was initially placed in front of a loudspeaker at a distance of 30 cm to test the effect of motor noise while it was walking and listening to the sound at the same time

remains (Figs. 6.8b and 6.9b, right). Most low amplitude noise (e.g., below 0) can be ignored and some part having high amplitudes (e.g., above 0) can be eliminated by the following network, called a sound–direction detection network (demonstrated later).

The second attempt was to observe the behavior of the evolved advanced auditory network when the signal having different waveforms was applied. Three waveforms were employed: sine, square and triangle shapes. All waveforms were generated at the same frequency—200 Hz—via a function generator. The signal from the function generator was directly connected to the analog port and digitized via the ADC channel of the MBoard. The digital signal was then provided as an input to the network. The input of the different waveforms, the FFT spectrum of each and the corresponding output of the network are shown in Fig. 6.10.

By testing with three different waveforms, the network apparently had a difficult time recognizing a triangle shape although it contains low-frequency signal (200 Hz). However, the network can obviously detect the signal of sine and square shapes even though the square shape is composed of more than one frequency (Fig. 6.10b, right). Thus, it can be concluded that not only the frequency but also the waveform of an input signal play important roles in the signal detection; i.e., the network can recognize the signal having sine and square waveforms at low frequencies while it cannot recognize the signal having triangle waveforms. Nevertheless, these network characteristics would be adequate for our approach, where the aim is to detect a sine wave signal to activate the sound tropism of the walking machine.

The last attempt was to show the performance of the sound–direction detection network (Fig. 5.11a) in filtering unwanted noise and discerning the direction of the signals. The stereo input given to the network was first filtered by the evolved advanced auditory network. However, there is remaining noise

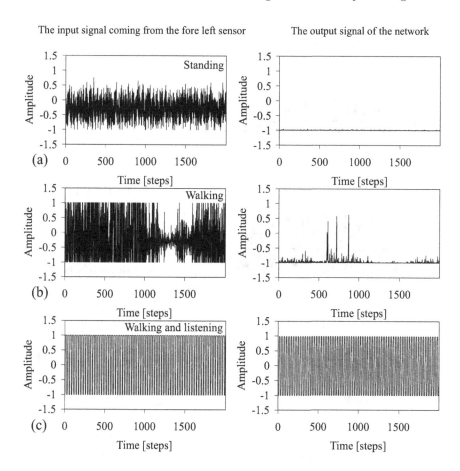

Fig. 6.8. *Left*: The input signal coming from the fore left sensor under three different conditions. *Right*: The output signal of the evolved advanced auditory network with respect to the input on the left side. (**a**) The noisy signal when the machine was in a standing position. (**b**) The noisy signal during the walk. (**c**) The noisy signal which was compensated between the sound and a noise during the walk. All figures have the same scale in the x-axis and the y-axis

which occurs from the locomotion of the walking machine. The network has to get rid of such noise and discern the direction of the stereo input signal based on the concept of the TDOA between left and right inputs. The capability of the sound–direction detection network in filtering the remaining noise is presented in Fig. 6.11.

The outputs of the sound–direction detection network (Figs. 6.11a and 6.11b, right) show that the network is able to filter the remaining noise which

The input signal coming from the rear right sensor   The output signal of the network

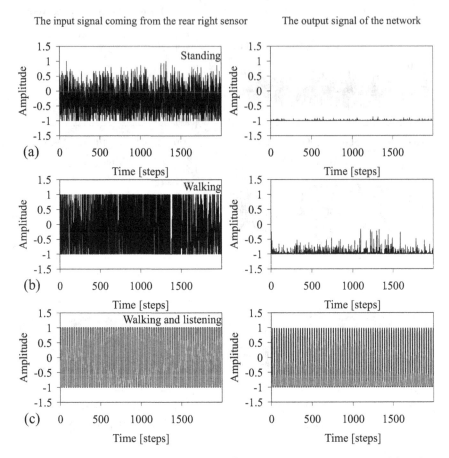

Fig. 6.9. *Left*: The input signal coming from the rear right sensor under three different conditions. *Right*: The output signal of the evolved advanced auditory network with respect to the input on the left side. (**a**) The noisy signal when the machine was in a standing position. (**b**) The noisy signal during the walk. (**c**) The noisy signal which was compensated between the sound and a noise during the walk. All figures have the same scale in the x-axis and the y-axis

comes from the machine while walking. As a result, no existing noise will disturb the controller for generating the behavior of the walking machine.

To test the ability of the network to discern the direction of the sound source, the walking machine was placed in front of a loudspeaker at a distance of 30 cm (Fig. 6.7) and the low-frequency sound at 200 Hz, having a basic sine shape, was generated. Additionally, the walking machine was manually turned to the opposite side during the experiment. The examples of input and output signals of the network are drawn in Fig. 6.12.

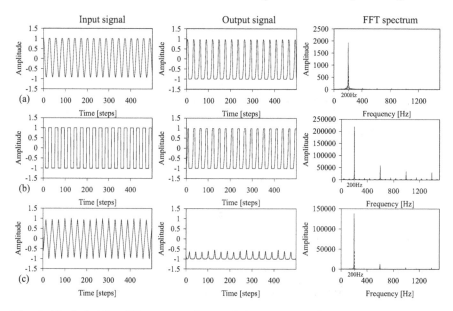

Fig. 6.10. *Left*: The different waveforms of the input signal at 200 Hz generated by a function generator. *Middle*: The output signal of the network with respect to the input signal on the left side. *Right*: The FFT spectrum of each input signal. (**a**) The signal having a sine shape. (**b**) The signal having a square shape. (**c**) The signal having a triangle shape

Figure 6.12 shows that the sound–direction detection network can distinguish the direction of the sound source by observing a leading signal or solely an active signal. In these example situations, when the signal of Input2 (I_2) leads the signal of Input1 (I_1) or only I_2 gets activated, this indicates that "the sound source is on the left" and the reverse case indicates that "the sound source is on the right".

6.1.3 The Antenna-like Sensor Data

In this section the tests for the preprocessing of the antenna-like sensor data (cf. Sect. 5.1.3) are presented. The IR-based antenna sensors together with the preprocessing shall be implemented on the walking machines for an obstacle avoidance task.

The following experiments were performed on the mobile processor (the PDA together with the MBoard) of the walking machines. The sensory inputs were digitized via the ADC channels of the MBoard at the sampling rate of up to 5.7 kHz. The preprocessing network was applied on the PDA with an update frequency of 75 Hz, and the communication between the board and the PDA was done by an RS232 interface at 57.6 kbits/s.

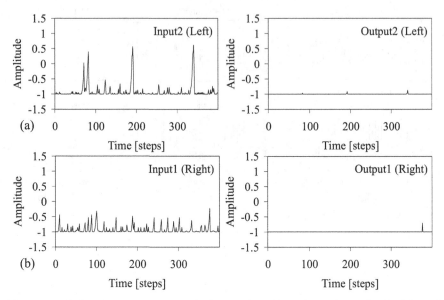

Fig. 6.11. *Left*: The input signals which were first filtered via the evolved advanced auditory network before going into the sound–direction detection network. *Right*: The output signals of the sound–direction detection network. (**a**), (**b**) The input and output signals of the sound–direction detection network on the *left* and *right*, respectively. Both unwanted parts of noise were almost removed by the sound–direction detection network

The experimental apparatus consists of two IR-based antenna sensors installed on a forehead of the walking machine AMOS-WD02, the mobile processor and the objects, which are boxes. The objects were placed in front of the walking machine at a distance of 25 cm for the left and right detectors. In order to observe the network behavior when both inputs are very highly activated, the objects were put at the closer distance of 10 cm. The experimental set-up is shown in Fig. 6.13.

Two networks for signal processing were introduced, a standard version working with two sensory inputs (compare with Fig. 5.20) and a developmental version working with more than two sensory inputs (compare with Fig. 5.21). However, only the performance of the standard version was shown in this experiment because both networks behave in the same manner. Three situations were carried out to provide the sensory information to the network (compare with Fig. 6.13). The sensory inputs from different situations are illustrated on the left of Fig. 6.14, and the resulting signals from the preprocessing network are shown on the right.

The preprocessing network functions as an on–off switch (Fig. 6.14); i.e., it switches on (Output neuron is active ($\approx +1$)) when the obstacles are detected;

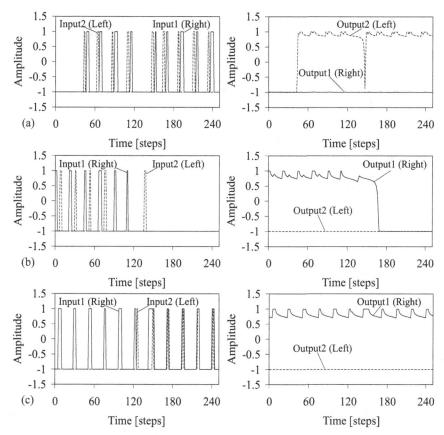

Fig. 6.12. *Left*: The input signals of the sound–direction detection network which were first filtered via the evolved advanced auditory network. *Right*: The output signals of the sound–direction detection network. (**a**) The signal of Input2 (I_2, *dashed line*) led the signal of Input1 (I_1, *solid line*) with the result that the signal of Output2 (O_2, *dashed line*) was active while the signal of Output1 (O_1, *solid line*) was inactive. The activated O_2 indicates that the sound source was on the left side. (**b**) I_2 followed I_1 with the delay resulting in O_2 being inactivated while O_1 was activated. The activated O_1 indicates that the sound source was on the right side. (**c**) In this situation, the network detected that the sound source was on the right side because I_1 was solely detected at the first period although I_2 having in phase with I_1 was also detected after around 150 time steps

otherwise it switches off (Output neuron is inactive (≈ -1)). This behavior of the network is mainly caused by the excitatory self-connection weights at the output neurons and the strong synapses from the input to the output units (Fig. 5.20). One of their properties, the noise of sensor data, is eliminated.

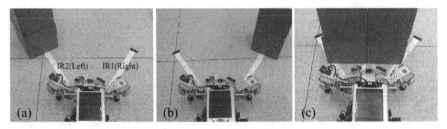

Fig. 6.13. (a) The situation where objects were presented on the left side in front of the walking machine at a distance of 25 cm. (b) The situation where objects were presented on the right side having the same distance like as (a). (c) The situation where objects were presented on both sides at the closer distance of 10 cm

The resulting smooth outputs together with the velocity regulating networks VRNs (cf. Sect. 5.2.2) will control the walking machines to avoid obstacles.

In some situations, like in a corner and in a deadlock, both input signals might be active. If both of them do not get a very high activation value, like the situation demonstrated in Fig. 6.14c, the network will provide only one active output ($\approx +1$) at a time. Such a situation was simulated and is shown in Fig. 6.15.

As a result, the network is able to control the output signals corresponding to the active input signals. Generally, only one output gets active at a time, which is determined by the previous active input. This phenomenon is mainly affected by an even loop between the output neurons of the network (see Sect. 5.1.3). By utilizing this effect to control the walking machines, they are then able to escape from a corner or a deadlock situation without getting stuck.

6.2 Implementation on the Walking Machines

In this section, the performance of the behavior controllers derived from the neural preprocessing and neural control is presented. The controllers, which generate the different reactive behaviors, were developed for a mobile system. The first attempt was to test the capability of the obstacle avoidance controller. After that the performance of the sound tropism controller was demonstrated, and the last attempt was to show the behavior fusion. It is controlled by the behavior fusion controller combination with the sensor fusion technique.

All the following experiments were performed on the four-legged walking machine AMOS-WD02 with installed physical sensor systems (the stereo auditory sensor and two antenna-like sensors) and all controllers were applied to the PDA. Additionally, the six-legged walking machine AMOS-WD06 with the

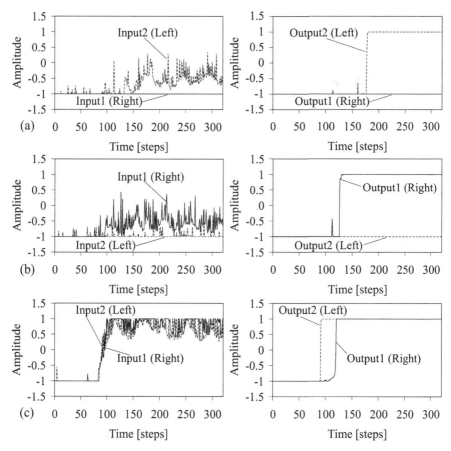

Fig. 6.14. (a) The situation where objects were fully presented on the left side after around 170 time steps. The left input signal (I_2, *dashed line*) was active after around that time causing the signal of Output2 (O_2, *dashed line*) to become active ($\approx +1$) while the signal of Output1 (O_1, *solid line*) remained inactive (≈ -1). (b) The situation where the objects were fully presented on the right side after around 120 time steps. The right input signal (I_1, *solid line*) was active after that time, causing O_1 to become active ($\approx +1$) while O_2 remained inactive (≈ -1). (c) The situation where the objects were presented on both sides. Although objects were presented on both sensors at the same time, I_2 was gradually activated to a high level and directly afterwards I_1 was activated to a high level following a similar pattern to I_2. Consequently, O_2 was activated first after around 90 time steps while O_1 became activated after around 120 time steps

installed six antenna-like sensors was also used to test the obstacle avoidance controller.

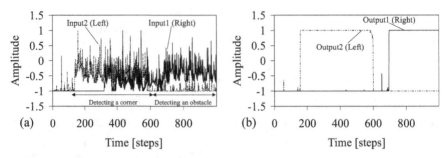

Fig. 6.15. (a)The input signals of the left (I_2, *dashed line*) and right (I_1, *solid line*) sensors. (b) The signals of Output1 (O_1, *solid line*) and Output2 (O_2, *dashed line*) correspond to the right and left inputs, respectively. At first, the left sensor detected one side of the corner after around 160 time steps while another side of the corner was also detected by the right sensor after around 300 time steps. Correspondingly, O_2 was excited ($\approx +1$) while O_1 was inhibited (≈ -1). After around 600 time steps, the left sensor did not detect the corner assuming that the machine had already turned right and then walked away from the corner. However, the right sensor was still active assuming that an obstacle was presented on the right side. This caused O_2 to become inactive and O_1 to become active

6.2.1 Obstacle Avoidance Behavior

This section describes experiments carried out to assess the ability of the obstacle avoidance controller to account for the obstacle behavior data. It focuses solely on avoiding obstacles, with the stereo auditory sensor system of the four-legged walking machine disabled at the sensor input level; i.e., the machine cannot react to any auditory signal in these experiments.

The performance of the obstacle avoidance controller (of the four- and six-legged walking machines) introduced in Sect. 5.3.1 was first tested in a simulated complex environment (cf. Sect. 4.2). It was then loaded into a mobile processor (the PDA) for a test on the physical autonomous walking machines.[2] However, the simulated walking machines and the physical walking machines behave similarly. The functionality and the property of the preprocessing of the antenna-like sensor data were shown in the section above. Here, the output signals of the network were directly connected to the neural control to modify the machine behavior as expected from a perception–action system. If obstacles are presented on either the right side or the left side, the controller will change the rhythmic movement of the legs at the thoracic joints, causing

---

[2] In the experiment, the AMOS-WD02 performs normal walking (without activating a backbone joint) with a walking cycle at 1.25 s or a walking speed at $\approx$ 0.45 body length/s (12.7 cm/s), while the AMOS-WD06 has a walking cycle at 1.52 s or a walking speed at $\approx$ 0.175 body length/s (7 cm/s). With these optimal walking speeds, the walking machines using battery packs can autonomously run up to 35 minutes during the experiments.

the walking machines to turn on the spot and immediately avoid the obstacles. In some situations, like approaching a corner or a deadlock situation, the preprocessing network determines the turning direction, left or right, with respect to the previously active input signal (see Sect. 6.1.3). The ability of the controller for the four-legged walking machine (cf. Fig. 5.37) which executes the obstacle avoidance behavior is illustrated in Fig. 6.16.

As shown in Fig. 6.16, Motor0 (M0) and Motor1 (M1) of the thoracic joints were turned to the opposite direction if the left sensor (IR2) detected the obstacle (compare a left column in Fig. 6.16). Correspondingly, Motor2 (M2) and Motor3 (M3) of the thoracic joints turned to the opposite direction when the right sensor (IR1) was active (compare a middle column in Fig. 6.16).

In special situations, e.g., walking toward the wall or detecting obstacles on both sides, both antenna-like sensors might be simultaneously active. Thus M0, M1, M2 and M3 of the thoracic joints turned to other directions which causes the walking machine to walk backward (compare a right column in Fig. 6.16). While walking backward one of the sensors might still be active, causing the active sensory signal to make the machine turn to the corresponding side until, eventually, it is able to leave the wall. Figure 6.17 displays a series of photos showing the avoidance of obstacles as well as the machine leaving from a deadlock situation.

The photos on the left column in Fig. 6.17 show that the walking machine can avoid an unknown obstacle, and it can also escape from a corner-like obstacle and a deadlock situation (see middle and right columns in Fig. 6.17). By using two antenna-like sensors installed at the forehead of the four-legged walking machine together with the neural controller, the walking machine has a capability to avoid the unknown obstacle as well as to escape from the corner or the deadlock situation.

However, some difficult situations were experienced in the presence of obstacles such as the legs of a chair or a desk. To protect the legs of the machine from colliding with these obstacles, more sensors need to be implemented on each leg of the machine, as mentioned in Chap. 4, and the preprocessing of the sensor data given in Sect. 5.1.3 is also required. The performance of the preprocessing for six sensory inputs combined with the neural control of the six-legged walking machine is exemplified in Figs. 6.18 and 6.19.

The modification of the signals at the motor neurons of the thoracic joints (M3, M4 and M5) into the reverse direction is shown in Fig. 6.18. There, object was presented to each of the right sensors at different time steps. Also, the controller changes the signals of the motor neurons (M0, M1 and M2) to the opposite direction when the object was presented to each of the left sensors at different time steps (Fig. 6.19). Figure 6.20 presents the example of reactive behavior of the walking machine AMOS-WD06 driven from the sensory inputs through the neural controller. A series of photos on the left and middle columns in Fig. 6.20 shows that the walking machine can protect its legs from colliding with the legs of the desk as well as the legs of the chair. Moreover, the walking machine was also able to turn away from the unknown

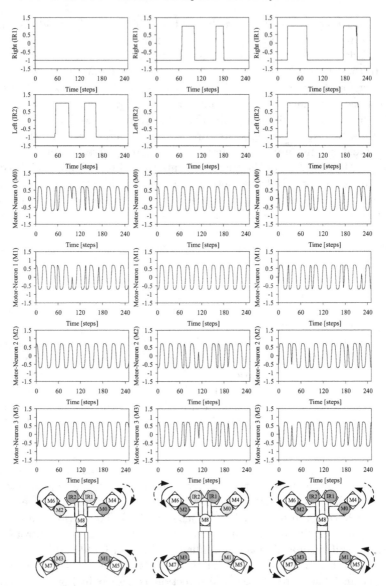

Fig. 6.16. *Left*: If the obstacles were presented on the left of the walking machine, then the output signals of the motor neurons (M0, M1) on its right change their direction as indicated by the *arrow dashed lines* in the *lower picture*. *Middle*: If the obstacles were detected on the right of the walking machine, then the motors (M2, M3) on its left would reverse as indicated by the *arrow dashed lines* in the *lower picture*. *Right*: In this situation, the obstacles were simultaneously detected on both sides resulting in the reversion of all motors (M0, M1, M2 and M3) as indicated by the *arrow dashed lines* in the *lower picture*, and the machine then walks backward

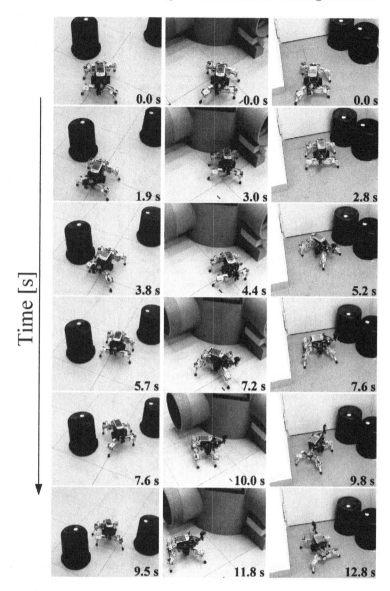

Fig. 6.17. Examples of the behavior driven by the antenna-like sensors of the four-legged walking machine AMOS-WD02. *Left*: The typical obstacle avoidance behavior. *Middle*: Another situation where the walking machine was able to avoid a corner. Comparing the two photos at 3.0 s and 4.4 s, one may observe that the machine is able to slightly step backward because both sensory signals were active at nearly the same time (at around 3.0 s). While walking backward (at around 4.4 s), the right sensor was still active while the left sensor was already inactive. Consequently, the walking machine turned left and walked away from the obstacle afterwards. *Right*: The walking machine was also able to escape from a deadlock situation without getting stuck

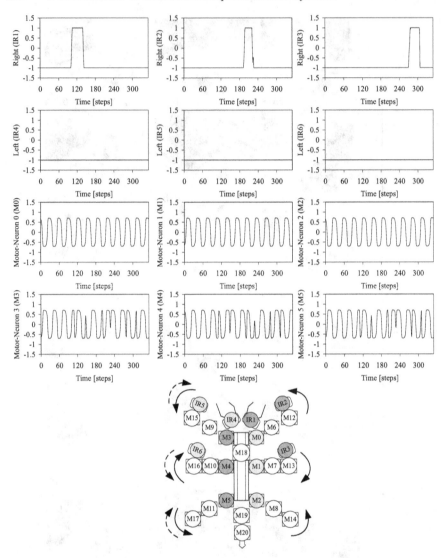

Fig. 6.18. An obstacle detected by each of the right sensors (IR1, IR2 and IR3) at different time steps; this caused the left motor neurons (M3, M4 and M5) to change into the opposite direction as indicated by the *arrow dashed lines* in the *lower picture*. As a result, the walking machine turns left

obstacles which were first sensed by the sensors at the forehead and then were detected by the sensors on the left legs (see right column in Fig. 6.20).

As demonstrated, the obstacle avoidance controller (of the four- and six-legged walking machines) is adequate to successfully solve the obstacle avoidance task. Additionally, the controller can protect the machines from getting

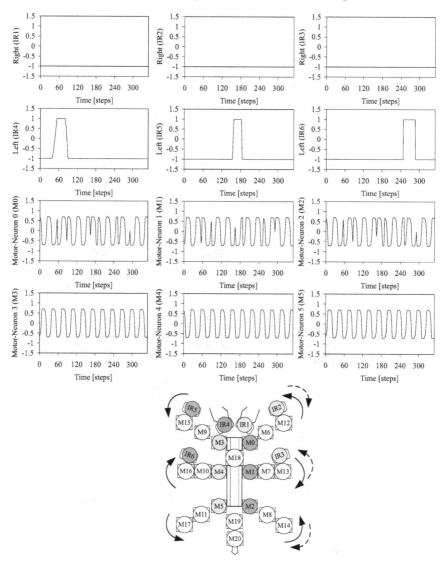

Fig. 6.19. An obstacle was presented at each of the left sensors (IR4, IR5 and IR6) at different time steps; this caused the right motor neurons (M0, M1 and M2) to change into the opposite direction as indicated by the *arrow dashed lines* in the *lower picture*. As a result, the walking machine turns right

stuck in the corner or the deadlock situation. Thus, due to this functionality, the reactive walking machines can automatically perform an exploration task or a wandering behavior.

Fig. 6.20. Examples of the behavior driven by the antenna-like sensors of the six-legged walking machine AMOS-WD06. *Left*: The walking machine was able to protect its legs from colliding with the leg of the desk which was detected by the sensors installed on the right legs of the machine. *Middle*: The machine was also able to avoid the legs of the chair. *Right*: The walking machine turned away from the unknown obstacles which were detected by the sensors at the forehead (IR1 and IR4) and then at the left legs (IR5 and IR6)

6.2.2 Sound Tropism

This section describes experiments carried out to test the capacity of the model performing the sound tropism. By now, it concerns solely a behavior reacting to an auditory signal. To do so, the stereo auditory sensor system was enabled while all antenna-like sensors were disabled at the sensor input level; i.e., the four-legged walking machine cannot avoid obstacles in these experiments.

The neural preprocessing of the stereo auditory signal was tested in the section above. The experimental results show that such preprocessing can filter unexpected noise. In addition, it can recognize and discern the direction of the sound source at low frequencies. Here, the combination of this preprocessing unit, called "auditory signal processing network", and the neural control unit leads to a so-called sound tropism controller (cf. Fig. 5.40). As a result, it performs a desired sound tropism in the four-legged walking machine AMOS-WD02.

The controller was applied to the PDA. It was then tested on the AMOS-WD02 in a real environment. An auditory signal having a sine shape at the frequency of 200 Hz was produced by a powered loudspeaker system (30 watts). The signal was detected via the stereo auditory sensor and was then digitized through the ADC channels of the MBoard at a sampling rate of up to 5.7 kHz.

For the first experiment, the maximum distance at which the system is able to detect the signal was measured. During the test, the signal was produced, and the experiment was repeated six times at each of the different locations shown in Fig. 6.21.

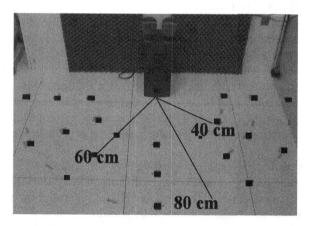

Fig. 6.21. The experimental set-up with the sound source and the markers (*black square areas*) where the walking machine was placed

The detection rates of the signal, i.e., the number of the correct detection[3] divided by the number of the experiments, are shown in Table 6.1. From the table, it can be concluded that the system can reliably react to the signal in the radius up to around 60 cm.

Table 6.1. Detection rate of the auditory signal at 200 Hz from different distances

Distance	Detection rate
40 cm	100%
60 cm	67%
80 cm	0%

The second experiment was to show the ability of the controller which can identify the location of the sound source (on the left or the right of the walking machine) and to present the modified signals of the motor neurons of the thoracic joints (M0, M1, M2 and M3). The sensory inputs coming from the right and left auditory sensors (Input1 and Input2, respectively) and the signals of the motor neurons were monitored. They are presented in Figs. 6.22 and 6.23.

As shown in Fig. 6.22, M2 and M3 of the thoracic joints turned to the opposite direction if the sound source was on the left of the walking machine. On the other hand, M0 and M1 of the thoracic joints would reverse to the other direction when the sound source was on its right (Fig. 6.23). Due to these effects, the controller has the capability to enable the walking machine to turn toward the sound source.

After the walking machine turns toward and approaches the sound source, it will stop (simulating that it is capturing its prey). This action can be performed by comparing the amplitude of either the left or right signal with a threshold value. That is, if and only if the amplitude of one signal is larger than a threshold value, then the signal of motor neurons (M0, M1, M2 and M3) will automatically be set to 0 with the result that the machine cannot turn left, right nor even step forward. The monitored amplitudes of the left and right signals together with the signals of the motor neurons are shown in Fig. 6.24.

The last task of this section was to display the sound tropism in the real environment. The walking machine started from different initial positions and the auditory signal of 200 Hz having a sine shape generated by a loudspeaker while the machine was walking. Figures 6.25 and 6.26 show a series of photos of these example experiments.

By observing the behavior of the walking machine in the given examples, one can see that the walking machine behaved almost the same. It turned

---

[3] Correct detection means that the machine can correctly discern if the signal is coming from the left or the right.

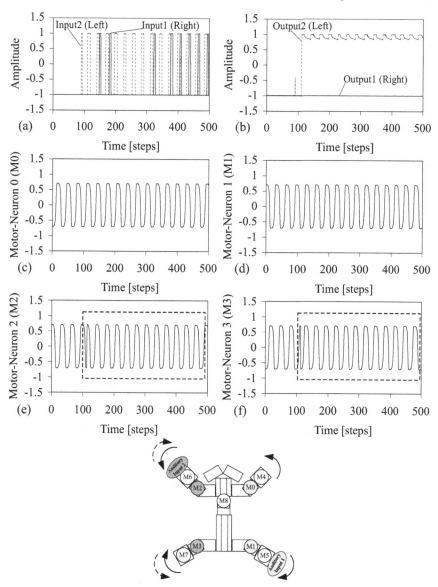

Fig. 6.22. (a) The auditory input signals from the left sensor (*dashed line*) and the right sensor (*solid line*) with the delay between each other. In this situation, Input2 led to Input1, indicating that the sound source was on the left of the walking machine. (b) The output signals after preprocessing via the auditory signal processing network. The network drove Output2 became activated while it inhibited Output1. (c), (d) The signals of the right motor neurons which are controlled by Output1 had no effect. (e), (f) The signals of the left motor neurons were modified (see *dashed frames*) because they are controlled by the activated Output2. The modified motors are also presented by the *arrow dashed lines* in the *lower picture*. Consequently, the walking machine turns left

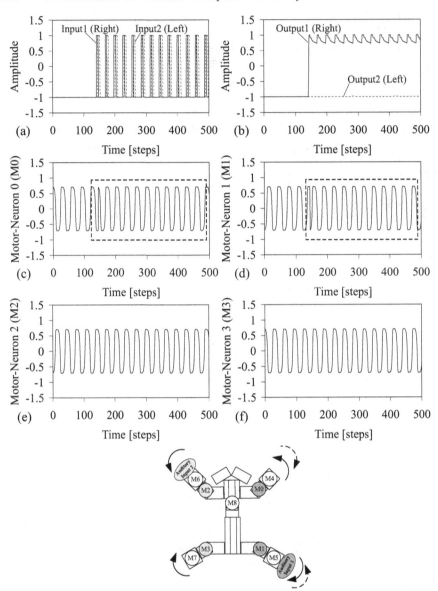

Fig. 6.23. (a) The auditory input signals from the right sensor (*solid line*) and the left sensor (*dashed line*) with the delay between each other. In this situation, Input1 led to Input2, indicating that the sound source was on the right. (b) The output signals after preprocessing via the auditory signal processing network. The network drove Output1 became activated while it inhibited Output2. (c), (d) Due to the activation of Output1, two motor neurons of the right thoracic joints were reversed (see *dashed frames*) while the signals of left motor neurons in (e) and (f) had no effect. The *arrow dashed lines* in the *lower picture* also show the reversion of right motors. Consequently, the walking machine turns right

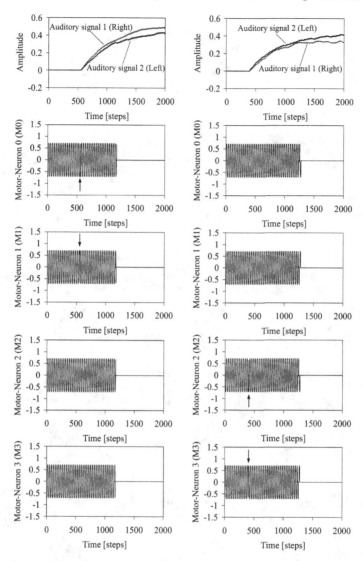

Fig. 6.24. *Left*: The sound source was on the right of the walking machine. During the first period, the walking machine was still far from the source and it walked closer to the source. After around 510 time steps the auditory signals were recognized and the signals of two motor neurons (M0, M1) were modified (indicated by *arrows*). The machine then turned right and approached the source; after around 1200 time steps the amplitude of the right signal was larger than the threshold value (here, 0.37). This results in the signals of the motor neurons (M0, M1, M2 and M3) being automatically set to 0. *Right*: In this situation, the sound source was on the left side. The auditory signals were detected after around 400 time steps and the signals of two motor neurons (M2, M3) were modified (indicated by *arrows*). Finally, the signals of the motor neurons (M0, M1, M2 and M3) were set to 0 after around 1300 time steps because the amplitude of the left signal was larger than 0.37

Fig. 6.25. The example of the sound tropism whereby the sound source was initially on the left of the walking machine and was generated during the walk. At the beginning, it walked forward and then it started to turn left at around 3.9 s because it detected the sound. Subsequently, it was steered to start turning right at around 12.5 s because the sound source now was on its right. Eventually, the machine made an approach and stopped in front of the source because the amplitude of the left sensor signal was higher than the threshold value

Fig. 6.26. The example of the sound tropism where the sound source was initially in front of the walking machine and was generated during the walk. In this situation, the walking machine also behaved like the previous example. It walked forward and then turned into the direction of the sound source when the sound was detected. After that, it approached and stopped beside the sound source

toward the sound source if it heard the sound; otherwise it kept walking forward until a threshold value was reached. This can be compared with one of the amplitude of the auditory signals. Finally, it stopped close to the sound source.

Additionally, the experimental results show that the walking machine has also oscillating-like movements when the sound is detected; i.e., it switched back and forth between turning left and right until it came close to the sound source. Also, it did not always reach the sound source with its head pointing to the source, but sometimes with the side of the body. However, these oscillating-like movements and approaching positions would not matter; if the walking machine reaches the sound source that is sufficient. In conclusion, the walking machine can successfully perform the sound tropism at the frequency of 200 Hz at the distance of up to around 60 cm.

6.2.3 Behavior Fusion

This last section will demonstrate behavior fusion between an obstacle avoidance including an exploration and a sound tropism in the four-legged walking machine. In this situation, all sensors were activated to sense the surrounding environment, e.g., detecting the obstacles via two antenna-like sensors and listening to sound via a stereo auditory sensor. The behavior fusion controller collaborating with a sensor fusion technique (see also Sect. 5.3.3) was employed for the behavior fusion approach. A part of the controller was implemented on the PDA while the other was programmed on the servo controller board. The sensor inputs were digitized via the ADC channels of the MBoard at a sampling rate of up to 5.7 kHz.

The controller will switch between two modes whereby one is called "obstacle avoidance mode (Om)", which enables the machine to solely avoid obstacles, and the other is known as the "composite mode (Cm)", which is capable of obstacle avoidance and sound detection. The executing time of each mode was optimized experimentally. Here, it was set to around 3.2 s for the obstacle avoidance mode and around 13.7 s for the composite mode; i.e., the obstacle avoidance mode will be first executed for around 3.2 s after that the composite mode will be executed for around 13.7 s. This process will be repeated until a processor time is reached, e.g., ≈ 15 minutes. However, one can remark that these desired executing and processor times can be adjusted depending on each physical perception–action system.

A series of photos in Figs. 6.27 and 6.28 presents the combined reactive behaviors of the walking machine AMOS-WD02 which can avoid the obstacles, wander around and also respond to a switched-on sound source when it can detect it. It is indicated at a lower left corner of each photo whether or not the sound source was switched on (On) or off (Off). In addition, the executed mode (Om or Cm) can be observed in the upper left corner and the action time is also shown in the lower right corner of each photo.

As demonstrated in Figs. 6.27 and 6.28, the behavior fusion controller together with the sensor fusion technique has the ability to generate different walking patterns which were driven by the sensory inputs, such that the machine could walk straight if no obstacle and no sound were detected. And then it turned toward a switched-on sound source and afterwards it would again continue walking forward without making the oscillating-like movement if and only if the obstacle avoidance mode is executed and no obstacle is detected. Eventually, it will approach and stop close to the sound source by determining one of the amplitudes of the auditory signals.

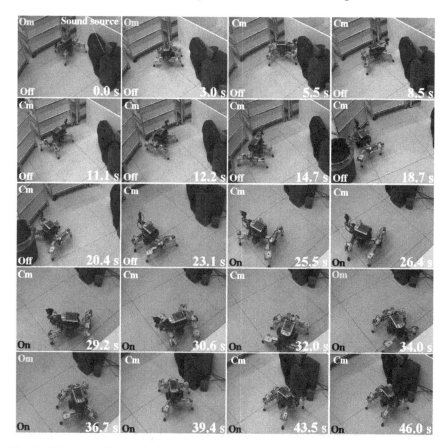

Fig. 6.27. At the first period, the source was switched off and the walking machine was wandering around and avoiding obstacles if they were detected. The source was then switched on at around 25.5 s to steer the walking machine; consequently, the machine started to turn left until around 34 s, at which point the obstacle avoidance mode was executed. Due to operating in the obstacle avoidance mode, and no obstacles being detected, the machine walked forward and got close to the source. Again the composite mode became activated at around 39.4 s, which made the walking machine turn slightly left and stop nearby the source at the end because the amplitude of the sensor signal was larger than the threshold value

However, there was a circumstance found in Fig. 6.28 at around 23.9 s. The walking machine turned right while it should normally turn left because the auditory signal of the left sensor was active. This occurred because the obstacle (a loudspeaker) was also detected at the same time, causing the IR signal of the left antenna-like sensor to be activated. Subsequently, both active signals (the left auditory signal and the left IR signal) were managed by the

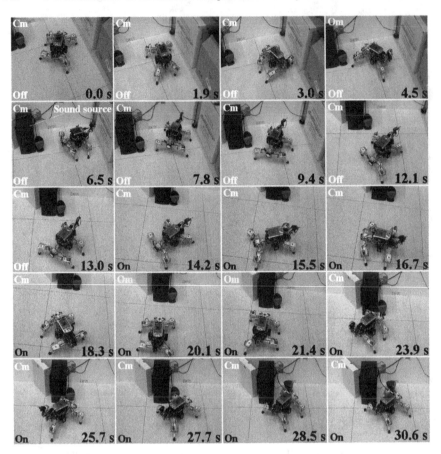

Fig. 6.28. The source was switched off and the walking machine was wandering around and avoiding obstacles if they were detected at the beginning. Then the source was switched on to control the walking machine to start turning right at around 14.2 s. Afterwards the walking machine started to walk forward at around 20.1 s because the obstacle avoidance mode was executed. It continued to walk forward until around 23.9 s, it turned right, not because of the sound but because of a detected obstacle instead (a loudspeaker), although the composite mode was operated. Eventually, the walking machine approached and stopped beside the source

sensor fusion technique in the composite mode (cf. Sect. 5.3.3). As a result, the observed behavior was performed.

6.3 Conclusion

The results given in this chapter showed that the neural preprocessing of the physical auditory–tactile sensor data has the ability to recognize two different signals with different frequencies that come from the tactile and auditory channels of the sensor. Also, the neural preprocessing of the stereo auditory sensor was tested with the real signal. It eliminates unexpected noise occurring from the motor sound of the walking machine, lets a low-frequency sound pass, and discerns the direction of the sound source (on the left or the right of the walking machine). Furthermore, the performance of the neural preprocessing of antenna-like sensor data was presented. It removes the noise of the sensor data and behaves like an on–off switch; i.e., it switches on (Output neuron is active) when the obstacles are detected otherwise it switches off (Output neuron is inactive).

The final section highlighted the co-operation between the different neural preprocessing units and the neural control unit leading to the behavior controllers to generate different reactive behaviors of the walking machine(s). First, the obstacle avoidance controller was implemented and tested on the physical walking machines (AMOS-WD02 and -WD06). They were able to avoid unknown obstacles and escape from a corner or a deadlock situation. One of them (AMOS-WD06), having more sensors installed on the two front and two middle legs, can ensure that its legs do not collide with obstacles, e.g., the legs of a desk or a chair. Second, the sound tropism was reproduced on the AMOS-WD02 by employing the sound tropism controller. It enables the walking machine to recognize the auditory signals coming from the left or the right. The machine turned into the direction of the sound source, then approached it, and finally stopped beside the source at the distance determined by a threshold of the amplitude of the signal. In the final demonstration, both reactive behaviors are fused and performed on the AMOS-WD02 by applying a so-called behavior fusion controller, which includes the sensor fusion technique. It generates a desired behavior driven by both auditory and IR stimuli. As a result, the walking machine wanders around, avoids obstacles, and walks toward and stops in front of the auditory signal (sound tropism).

7

Conclusions

7.1 Summary of Contributions

This book presents biologically inspired walking machines (four- and six-legged walking machines) interacting with their real environmental stimuli as agent–environment interactions. Different reactive behaviors of animals were investigated for the behavior design of the walking machine(s). On the one hand, the obstacle avoidance behavior, in analogy to the obstacle avoidance and escape behavior of scorpions and cockroaches, was implemented in the walking machines as a negative tropism. On the other hand, the sound tropism which mimics prey capture behavior of spiders is represented as a positive tropism. It was simulated on the four-legged walking machine.

The biological sensing systems which are used to trigger the described reactive behaviors were also investigated. Three types of sensory systems, which are an auditory–tactile sensor, a stereo auditory sensor and antenna-like sensors were constructed with respect to the biological sensing systems. The auditory–tactile sensor, which was inspired by the function of hairs of a scorpion and a spider, is used for tactile sensing as well as sound detection. Using the stereo auditory sensor in analogy to the hairs of the spider, the sound can be detected and the direction of the incoming sound can also be distinguished by determining the TDOA from the left and right auditory sensors. The antenna-like sensors, which were modeled with respect to the basic function of insect antennas, are used to detect impediments as well as to protect the legs of the six-legged walking machine from colliding with obstacles.

In addition, the morphologies of a salamander and a cockroach, which are used to perform efficient locomotion, were also considered for the leg and trunk designs of the four- and six-legged walking machines, respectively. They were successfully built with mechanical constructions. The rhythmic movements of the legs of the machines are basically generated by the CPG which corresponds to the basic locomotion control of walking animals.

The main focus of this book is not only to generate biologically inspired reactive behaviors in the physical walking machine(s) but also to present the simple mechanism for the desired behavior controls. On the basis of a modular neural structure, they were built from a combination of different neural preprocessing units for sensor data processing and the neural control unit for locomotion control of the walking machines. This means that each neural preprocessing unit can be connected with a neural control unit to obtain a different behavior control. Neural preprocessing and control were achieved by applying the discrete-time dynamical properties of recurrent neural networks generated by the evolutionary algorithm ENS^3. Three types of neural preprocessing were presented: auditory signal processing, preprocessing of the antenna-like sensor data and tactile signal processing. Auditory signal processing is used to recognize the low-frequency sound at 200 Hz for producing the sound tropism while it filters background noise at high frequencies ($>$400 Hz). In other words, it acts as a simple low-pass filter with its cutoff frequency at approximately 400 Hz. It also has the capability to discern the direction of the auditory signals coming from either the left or the right. For the preprocessing of the antenna-like sensor data, it is able to eliminate the sensory noise and to control an obstacle avoidance behavior. Applying the auditory–tactile sensor for collision detection and low-frequency sound detection, the signal coming from the tactile channel is recognized by the tactile signal processing network while the low-frequency sound is recognized by a part of the auditory signal processing called "the advanced auditory network".

The neural control was formed with two subordinate neural networks: the neural oscillator network, which generates the rhythmic leg movements as a central pattern generator, and the VRNs, which expand the steering capabilities of the walking machines. This neural control was created for generating a typical trot gait of the four-legged walking machine. Then it was modified (still having the same structure except more output motor neurons) to move the six-legged walking machine with a typical tripod gait.

Eventually, the integration between neural control and different types of neural preprocessing leads to several behavior controllers. For example, the obstacle avoidance controller is formed by connecting the preprocessing of the antenna-like sensor data with the neural control while the sound tropism controller is constructed by replacing preprocessing of the antenna-like sensor data with the auditory signal processing. Furthermore, a sensor fusion technique was employed. It combines all preprocessed sensory signals of the preprocessing of the antenna-like sensor data and the auditory signal processing to obtain a so-called behavior fusion controller.

Three behavior controllers together with the associated sensory systems were successfully implemented and tested on the walking machine(s). First, the obstacle avoidance controller was implemented on the physical walking machines. The walking machines were able to avoid unknown obstacles and escape from a corner or a deadlock situation. Moreover, one of them (the six-legged walking machine), having more sensors installed on the two front and two middle legs, can ensure that its legs do not collide with obstacles, e.g., the legs of a desk or a chair. Second, the sound tropism was reproduced on the four-legged walking machine by employing the sound tropism controller. It enables the walking machine to recognize the auditory signals (sinusoidal sound at 200 Hz) coming from the left or the right at a distance of up to approximately 60 cm. The machine turned toward the source like a predator reacting to a prey signal, then approached it, and finally stopped beside the source at the distance determined by a threshold of the amplitude of the signal (simulating that it is capturing its prey). In the final demonstration, both reactive behaviors are combined to one controller and then implemented on the four-legged walking machine. It generates the desired behavior, i.e., positive and negative tropism. The walking machine, as a result, reacts to the auditory signal, wanders around, avoids obstacles and even escapes from corners as well as deadlock situations.

The resulting reactive behaviors of the physical embodied system(s) show that the behavior controllers are robust and sufficient to deal with real unexpected noise, and due to the modular neural structure they are then flexible to adapt to the various target systems with a different complexity. On the other hand, they prove that the discrete-time dynamical properties of recurrent neural networks (e.g., hysteresis effect) together with an evolutionary algorithm can be applied to find the appropriate solution for neural preprocessing and control in a robotic domain. The described systems can also be defined as versatile artificial perception–action systems; i.e., they perceive environmental stimuli and display the corresponding actions without knowledge of an environmental model.

7.2 Possible Future Work

The work presented in this book was intended to be a basic step to achieve an "Autonomous Intelligent System", which should maintain its energy supply, survive in complex environments, show a certain degree of autonomy (although no robotic system is totally autonomous), learn to behave in an efficient way, etc. Thus, possible work based on the existing systems may be extended to:

- add an additional sensor like an energy sensor to monitor the energy consumption and to activate efficient walking gaits in a specific condition for maintaining the energy supply;

- add proprioceptors like foot contact sensors for ground sensing and angle encoders of joints to detect the movement of the legs and so on;

- implement more reactive behaviors (e.g., avoiding a predatory attack, phototropism) and using an evolutionary algorithm to cooperate or complete all these different reactive behaviors;

- utilize a learning technique, e.g., reinforcement learning, to allow the walking machines to behave in an efficient way (e.g., learning to find the fastest way to escape from an undesired situation or to make an approach to a target).

However, it would also be interesting to enable the walking machine to interact not only with its environment but also with other machines (agent–agent interactions). On the one hand, one may think about predator–prey interactions. While the walking machines can also assist each other when the requested signal is perceived. Generally, robotic models are indeed suited to the investigation of how behavior decisions arise from multiple sources of sensory information and can establish these ideas in specific neural mechanisms. Moreover, they can be used as tools to establish the relationship between biology, (computational) neuroscience and engineering as was shown in this book.

A

Description of the Reactive Walking Machines

Two physical reactive walking machines were built as mobile robot platforms. They are used for experiments with neural controllers so as to perform different reactive behaviors and also to demonstrate artificial perception–action systems.

A.1 The AMOS-WD02

The AMOS-WD02 is a four-legged walking machine with two degrees of freedom by leg. Its body consists of an (active) tail and a central chassis, which is connected to its head through an (active) backbone joint rotating on a vertical axis. Two rear legs are attached at the central chassis while another two are fixed at the head (Fig. A.1).

Fig. A.1. The physical four-legged walking machine AMOS-WD02. *Left*: *Top view* while turning its backbone joint. *Right*: *Front view* while in its standing position

Some basic characteristics that define the AMOS-WD02 are the following:

Mechanics

- Dimensions without the tail (L x B x H): 28 x 30 x 14 cm.
- Weight: 3.3 kg.
- Structure of polyvinyl chloride (PVC) and aluminum alloys AL5083.
- Four legs with two degrees of freedom in each leg.
- Backbone joint rotating on a vertical axis.
- Active tail rotating on horizontal and vertical axis.
- Driven by eight analog (90 Ncm), one digital (220 Ncm) and two micro (20 Ncm) servomotors.

Electronics

- Multi-Servo IO-Board (MBoard)[1] developed at the Fraunhofer Institute in Sankt Augustin. It is able to control up to 32 servomotors simultaneously. At the same time, 32 (+4 optional) analog input channels can be sampled and read with an update rate of up to 50 cycles per second. The board has an RS232 interface, which serves as the standard communication interface.
- Personal digital assistant (PDA) having an Intel (R) PXA255 processor for programming neural preprocessing and control. It communicates with the MBoard via an RS232 interface.
- The support circuitry of the auditory sensors.
- Battery of 4.8 V NiMH 2100 mAh for the servomotors.
- Battery of 4.8 V NiMH 2100 mAh for the support circuitry of the auditory sensors.
- Battery of 9 V NiMH for the MBoard.
- Battery of 9 V NiMH for the wireless camera.
- Two distance-measurement infrared sensors (antenna-like sensors) located at the forehead.
- Two auditory sensors located at the fore left and rear right legs.
- Mini wireless camera built in a microphone installed on the top of the tail.

Programming

- C programming on the MBoard for controlling servomotors and for reading digitized sensor data.
- Embedded Visual C++ on the PDA for programming neural preprocessing and control.

A.2 The AMOS-WD06

The AMOS-WD06 is a six-legged walking machine with three degrees of freedom by leg. Its body consists of an (active) tail and a central chassis which

---

[1] See also: http://www.ais.fraunhofer.de/BE/volksbot/mboard.html. Cited 18 December 2005.

is connected to its head through an (active) backbone joint rotating on a horizontal axis. Two legs are attached at the rear of the central chassis, and another two are installed at the front of the central chassis while the rest are fixed at the head (Fig. A.2).

Fig. A.2. The physical six-legged walking machine AMOS-WD06. *Left*: *Top view* in its climbing position. *Right*: *Front view* in its standing position

Some basic characteristics that define the AMOS-WD06 are the following:

Mechanics

- Dimensions without the tail (L x B x H): 40 x 30 x 12 cm.
- Weight: 4.2 kg.
- Structure of polyvinyl chloride (PVC) and aluminum alloys AL5083.
- Six legs with three degrees of freedom in each leg.
- Backbone joint rotating on a horizontal axis.
- Active tail rotating on horizontal and vertical axis.
- Driven by eighteen analog (100 Ncm), one digital (220 Ncm) and two micro (20 Ncm) servomotors.

Electronics

- MBoard which is able to control up to 32 servomotors simultaneously. At the same time, 32 (+4 optional) analog input channels can be sampled and read with an update rate of up to 50 cycles per second. The board has an RS232 interface, which serves as the standard communication interface.
- PDA having an Intel (R) PXA255 processor for programming neural pre-processing and control. It communicates with the MBoard via an RS232 interface.
- Battery of 6 V NiMH 3600 mAh for the servomotors.
- Battery of 4.8 V NiMH 800 mAh for six distance measurement infrared sensors.
- Battery of 9 V NiMH for the MBoard.

- Battery of 9 V NiMH for a wireless camera.
- Six distance-measurement infrared sensors (antenna-like sensors) where two of them were located at the forehead while the rest of them were fixed at the levers of the two front and two middle legs.
- Mini wireless camera built in a microphone installed on the top of the tail.
- One upside-down detector located beside of the body.

Programming

- C programming on the MBoard for controlling servomotors and for reading digitized sensor data.
- Embedded Visual C++ on the PDA for programming neural preprocessing and control.

A.3 Mechanical Drawings of Servomotor Modules and the Walking Machines

The drawings of a set of joint modules for the digital and analog servomotors and the walking machine constructions which were manufactured by aluminum alloys are shown in Figs. A.3–A.12.

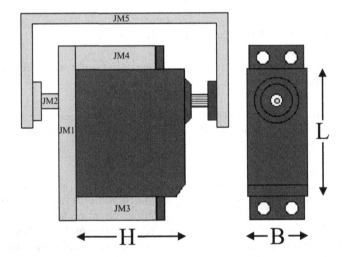

Fig. A.3. The drawing of a set of joint modules ($JM_{1,2,3,4,5}$) for the digital and analog servomotors. The size of the servomotor (L x B x H): 40.5 x 20 x 40.5 mm with a weight of 65 g

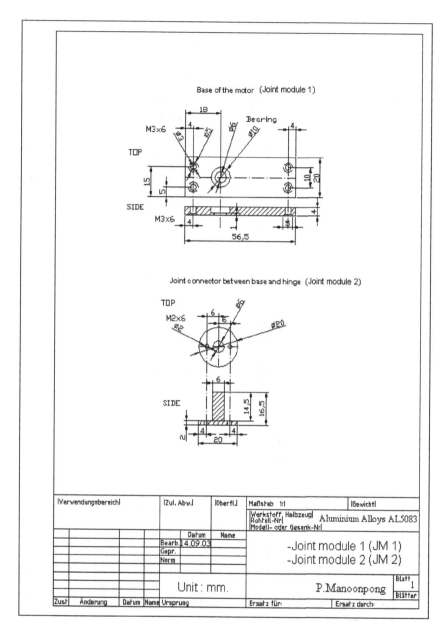

Fig. A.4. The drawing of the joint modules 1 and 2 ($JM_{1,2}$) of the servo motor

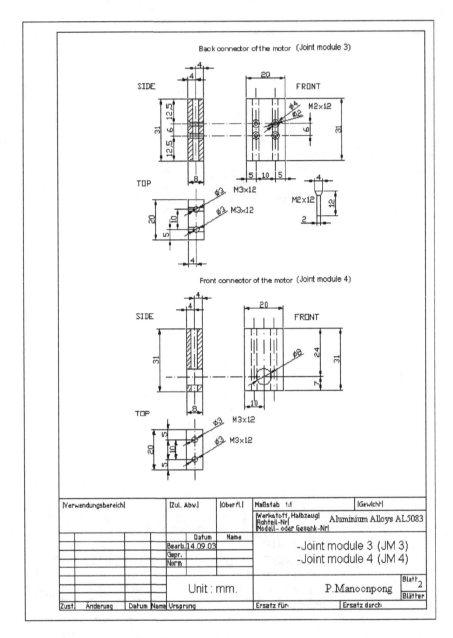

Fig. A.5. The drawing of the joint modules 3 and 4 ($JM_{3,4}$) of the servo motor

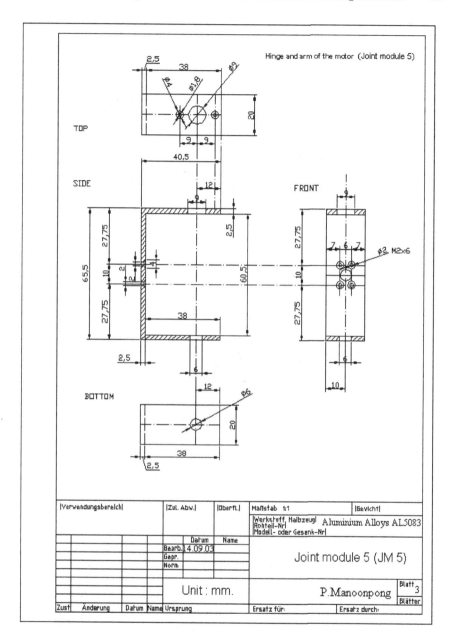

Fig. A.6. The drawing of the joint module 5 (JM_5) of the servo motor

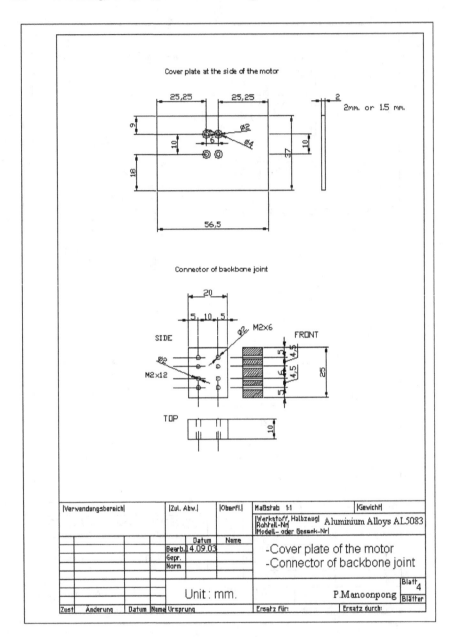

Fig. A.7. The drawing of a cover plate at each side of the servo motor and a backbone joint connector

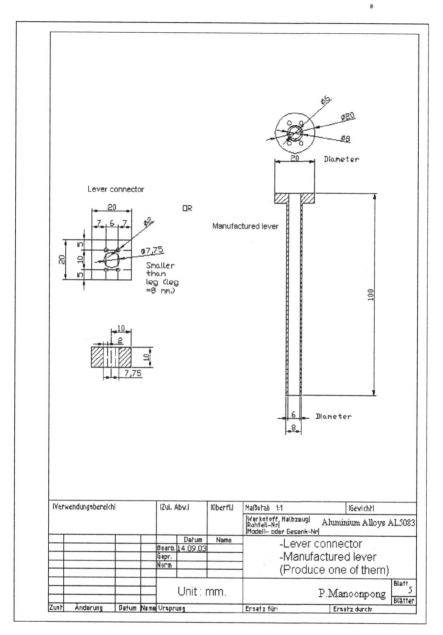

Fig. A.8. The drawing of a lever and its connector

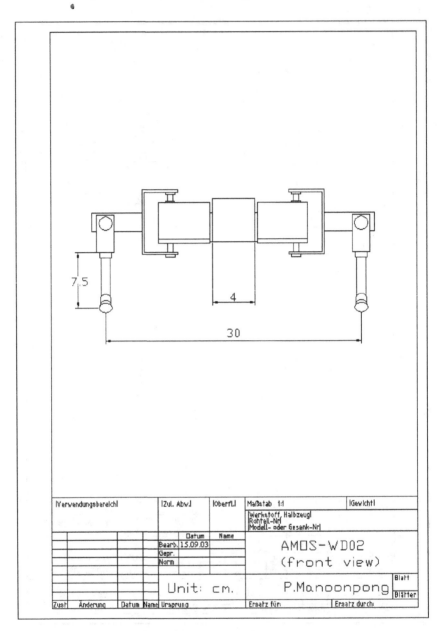

Fig. A.9. The drawing of the AMOS-WD02 (*front view*)

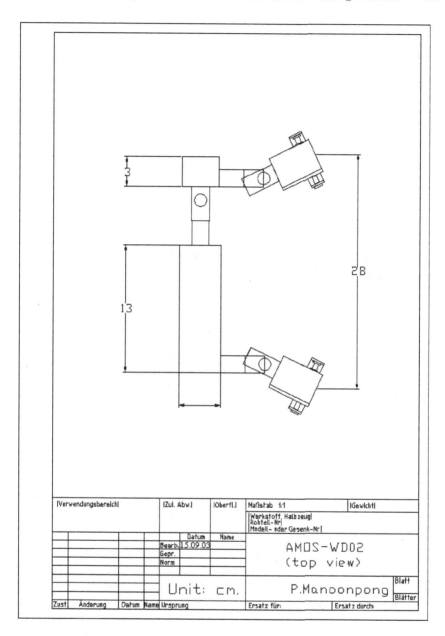

Fig. A.10. The drawing of the AMOS-WD02 (*top view*)

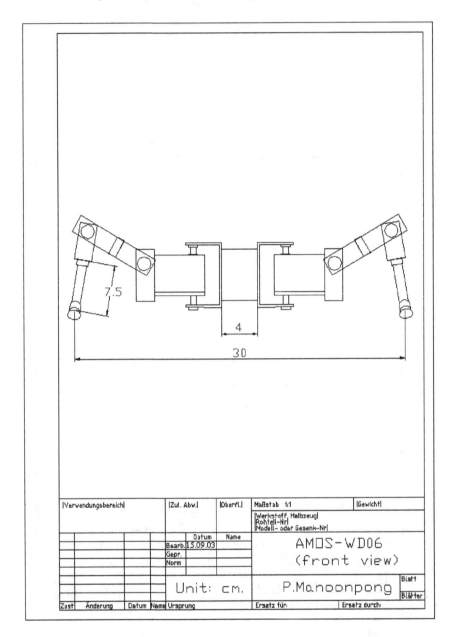

Fig. A.11. The drawing of the AMOS-WD06 (*front view*)

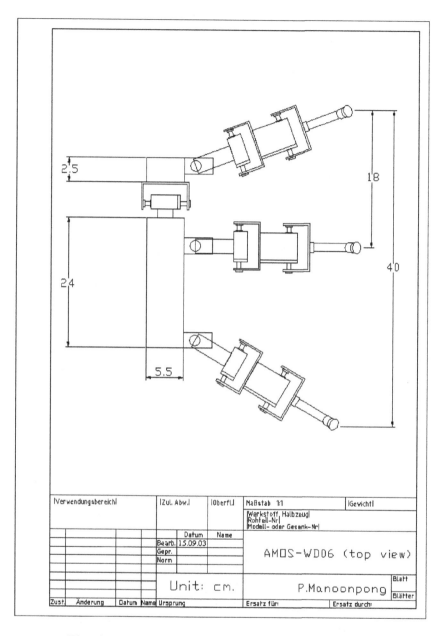

Fig. A.12. The drawing of the AMOS-WD06 (*top view*)

B

Symbols and Acronyms

List of Symbols

a_i	the activity of neuron i
b_i, B_i	a fixed internal bias term
Cm	the composite mode
E	the mean squared error
F	the fitness function
I_i	the input of neuron i
M_n	the motor neuron n
N	the maximal number of time steps
$o_i, O_i = f(a_i)$	the output of neuron i
Om	the obstacle avoidance mode
w_{ij}, W_{ij}	the synaptic strength of the connection from neuron j to neuron i
θ	the sum of a fixed internal bias term and the variable total input I of the neuron

List of Acronyms

ADC	Analog to Digital Converter
AL	Auditory signal of the Left auditory sensor
AMOS-WD	Advanced MObility Sensor driven-Walking Device
AMOS-WD02	The four-legged walking machine
AMOS-WD06	The six-legged walking machine
ANNs	Artificial Neural Networks
AR	Auditory signal of the Right auditory sensor
CPG	Central Pattern Generator
DOF	Degrees Of Freedom
ENS^3	Evolution of Neural Systems by Stochastic Synthesis
FFT	Fast Fourier Transform
IR	Infrared
IRL	Infrared signal of the Left antenna-like sensor
IRR	Infrared signal of the Right antenna-like sensor
ISEE	Integrated Structure Evolution Environment
MBoard	Multi-Servo IO-Board
MERLIN	Mobile Experimental Robots for Locomotion and Intelligent Navigation
MRC	Minimal Recurrent Controller
ODE	Open Dynamics Engine
PC	Personal Computer
PDA	Personal Digital Assistant
PMD	Photonic Mixer Device
PWM	Pulse Width Modulation
RNNs	Recurrent Neural Networks
TDOA	Time Delay Of Arrival
VRN	Velocity Regulating Network
YARS	Yet Another Robot Simulator

References

[1] (2002). Australian Museum. http://www.amonline.net.au. Cited 4 December 2005

[2] (2004). Yet Another Robot Simulator (YARS). http://www.ais.fraunhofer.de/INDY/, see menu item TOOLS. Cited 18 December 2005

[3] Aarabi, P.; Wang, Q. H.; Yeganegi, M. (2004). Integrated displacement tracking and sound localization. In: *Proceedings of the IEEE International Conference on Acoustics, Speech, and Signal Processing (ICASSP'04), vol. 5*, pp. 937–940

[4] Abushama, F. T. (1964). On the behaviour and sensory physiology of the scorpion *Leirus quinquestriatus. Animal Behaviour* **12(1)**, 140–153

[5] Albiez, J. C.; Luksch, T.; Berns, K.; Dillmann, R. (2003). Reactive reflex based control for a four-legged walking machine. *Robotics and Autonomous Systems* **44(3)**, 181–189

[6] Ali, K. S.; Arkin, R. C. (1998). Implementing schema-theoretic models of animal behavior in robotic systems. In: *Proceedings of the Fifth International Workshop on Advanced Motion Control (AMC'98)*, pp. 246–253

[7] Anderson, J. A. (1995). *An Introduction to Neural Networks*. Cambridge, Massachusetts: MIT Press

[8] Anderson, T. L.; Donath, M. (1988). A computational structure for enforcing reactive behavior in a mobile robot. In: W. J. William (ed.), *Proceedings of the SPIE Conference on Mobile Robots III*, Cambridge, Massachusetts, *vol. 1007*, p. 370

[9] Arbib, M. A. (1964). *Brains, Machines and Mathematics*. New York: McGraw-Hill

[10] Arkin, R. C. (1998). *Behavior-Based Robotics*. Cambridge, Massachusetts: MIT Press

[11] Ayers, J.; Davis, J.; Rudolph, A. (eds.) (2002). *Neurotechnology for Biomimetic Robots*. Cambridge, Massachusetts: MIT Press

[12] Ayers, J.; Witting, J.; McGruer, N.; Olcott, C.; Massa, D. (2000). Lobster robots. In: T. Wu; N. Kato (eds.), *Proceedings of the International*

Symposium on Aqua Biomechanisms, Hawaii: Tokai University Pacific
Center

[13] Azmy, N.; Boussard, E.; Vibert, J. F.; Pakdaman, K. (1996). Single neuron with recurrent excitation: Effect of the transmission delay. *Neural Networks* **9(5)**, 797–818

[14] Barreto, G. de A.; Araújo, A. F. R. (1999). Unsupervised learning and recall of temporal sequences: An application to robotics. *International Journal Neural System* **9(3)**, 235–242

[15] Barth, F. G. (2002). *A Spider's World: Senses and Behavior*. Berlin Heidelberg New York: Springer

[16] Barth, F. G.; Geethabali (1982). Spider vibration receptors: Threshold curves of individual slits in the metatarsal lyriform organ. *Journal of Comparative Physiology A: Neuroethology, Sensory, Neural, and Behavioral Physiology* **148(2)**, 175–185

[17] Bath, F. G.; Höller, A. (1999). Dynamics of arthropod filiform hairs. V. The response of spider trichobothria to natural stimuli. *Philosophical Transactions of the Royal Society of London Series B: Biological Sciences* **354**, 183–192

[18] Barth, F. G.; Humphrey, J. A. C.; Secomb, T. W. (eds.) (2003). *Sensors and Sensing in Biology and Engineering*. Berlin Heidelberg New York: Springer

[19] Barth, F. G.; Wastl, U.; Humphrey, J. A. C.; Devarakonda, R. (1993). Dynamics of arthropod filiform hairs-II: Mechanical properties of spider trichobothria (Cupiennius salei). *Philosophical Transactions of the Royal Society of London Series B: Biological Sciences* **340**, 445–461

[20] Bässler, U. (1983). *Neural Basis of Elementary Behavior in Stick Insects*. Berlin Heidelberg New York: Springer

[21] Bässler, U.; Büschges, A. (1998). Pattern generation for stick insect walking movements–multisensory control of a locomotor program. *Brain Research Reviews* **27**, 65–88

[22] Beer, R. D. (1990). *Intelligence as Adaptive Behavior: An Experiment in Computational Neuroethology*. New York: Academic

[23] Beer, R. D.; Chiel, H. J.; Quinn, R. D.; Espenschield, K. S.; Larsson, P. (1992). A distributed neural network architecture for hexapod robot locomotion. *Neural Computation* **4(3)**, 356–365

[24] Beer, R. D.; Chiel, H. J.; Sterling, L. S. (1990). A biological perspective on autonomous agent design. *Robotics and Autonomous Systems* **6(1–2)**, 169–186

[25] Beer, R. D.; Ritzmann, R. E.; McKenna, T. (eds.) (1993). *Biological Neural Networks in Invertebrate Neuroethology and Robotics (Neural Networks, Foundations to Applications)*. Boston, Massachusetts: Academic

[26] Bekey, G. A. (2005). *Autonomous Robots From Biological Inspiration to Implementation and Control*. Cambridge, Massachusetts: MIT Press

[27] Berns, K.; Cordes, S.; Ilg, W. (1994). Adaptive, neural control architecture for the walking machine LAURON. In: *Proceedings of the IEEE/RSJ*

International Conference on Intelligent Robots and Systems (IROS), vol. 2, pp. 1172–1177

[28] Berns, K.; Ilg, W.; Eckert, M.; Dillmann, R. (1998). Mechanical construction and computer architecture of the four-legged walking machine BISAM. In: *Proceedings of the First International Symposium on Climbing and Walking Robots (CLAWAR'98)*, pp. 167–172

[29] Billard, A.; Ijspeert, A. J. (2000). Biologically inspired neural controllers for motor control in a quadruped robot. In: *Proceedings of the IEEE-INNS-ENNS International Joint Conference on Neural Networks (IJCNN 2000)*, vol. 6, pp. 637–641

[30] Bongard, J.; Zykov, V.; Lipson, H. (2006). Resilient machines through continuous self-modeling. *Science* **314(5802)**, 1118–1121

[31] Böhm, H. (1995). Dynamic properties of orientation to turbulent air current by walking carrion beetles. *Journal of Experimental Biology* **198(9)**, 1995–2005

[32] Braitenberg, V. (1984). *Vehicles: Experiments in Synthetic Psychology.* Cambridge, Massachusetts: MIT Press

[33] Breithaupt, R. (2001). Walking–Robot–Kit: Modular and easy to enhance robot-kit for serious research and development. http://www.ais.fraunhofer.de/~breitha/projects/RoboKit/RoboKit.html. Cited 16 December 2005

[34] Breithaupt, R.; Dahnke, J.; Zahedi, K.; Hertzberg, J.; Pasemann, F. (2002). Robo-Salamander an approach for the benefit of both robotics and biology. In: P. Bedaud (ed.), *Proceedings of the Fifth International Conference on Climbing and Walking Robots (CLAWAR'02)*, London: Professional Engineering, pp. 55–62

[35] Brittinger, W. (1998). *Trichobothrien, Medienströmung und das Verhalten von Jagdspinnen (Cupiennius salei, Keys.).* Ph.D. thesis, University of Vienna, Austria. [Translation of title: Trichobothria, medium flow, and the behavior of hunting spiders (*Cupiennius salei Keys*)]

[36] Brooks, R. A. (1989). A robot that walks: Emergent behaviors from a carefully evolved network. *Neural Computation* **1(2)**, 253–262

[37] Brooks, R. A. (1991). How to build complete creatures rather than isolated cognitive simulators. In K. VanLehn (ed.), *Architectures for Intelligence*, pp. 225-239

[38] Brooks, R. A. (1999). *Cambrian Intelligence: The Early History of the New AI.* Cambridge, Massachusetts: MIT Press

[39] Brooks, R. A. (2002). *Flesh and Machines: How Robots Will Change Us.* New York: Pantheon Books

[40] Brooks, R. A.; Connell, J. H.; Ning, P. (1988). *Herbert: A Second Generation Mobile Robot.* Technical Report AI–Memo 1016, MIT Artificial Intelligence Laboratory

[41] Brooks, R. A.; Stein, L. A. (1994). Building brains for bodies. *Autonomous Robots* **1(1)**, 7–25

[42] Brown, T. G. (1911). The intrinsic factors in the act of progression in the mammal. *Proceedings of the Royal Society of London Series B* **84**, 308–319

[43] Buehler, M. (2002). Dynamic locomotion with one, four and six-legged robots. *Journal of the Robotics Society of Japan* **20(3)**, 15–20

[44] Büschges, A. (2005). Sensory control and organization of neural networks mediating coordination of multisegmental organs for locomotion. *Journal of Neurophysiology* **93**, 1127–1135

[45] Camhi, J. M.; Johnson, E. N. (1999). High-frequency steering maneuvers mediated by tactile cues: Antennal wall-following in the cockroach. *Journal of Experimental Biology* **202(5)**, 631–643

[46] Chapman, T. (2001). *Morphological and Neural Modelling of the Orthopteran Escape Response*. Ph.D. thesis, University of Stirling, UK

[47] Clark, J. E.; Cham, J. G.; Bailey, S. A.; Froehlich, E. M.; Nahata, P. K.; Full, R. J.; Cutkosky, M. R. (2001). Biomimetic design and fabrication of a hexapedal running robot. In: *Proceedings of the IEEE International Conference on Robotics and Automation (ICRA), vol. 4*, pp. 3643–3649

[48] Cloudsley-Thompson, J. L. (1958). *Spiders, Scorpions, Centipedes and Mites (The Ecology and Natural History of Woodlice, Myriapods and Arachnids)*. New York: Pergamon

[49] Comer, C. M.; Dowd, J. P. (1992). Multisensory processing for movement: Antennal and cercal mediation of escape turning in the cockroach. In: R. D. Beer; R. E. Ritzmann; T. McKenna (eds.), *Biological Neural Networks in Invertebrate Neuroethology and Robotics (Neural Networks, Foundations to Applications)*, Boston, Massachusetts: Academic, pp. 89–112

[50] Comer, C. M.; Mara, E.; Murphy, K. A.; Getman, M.; Mungy, M. C. (1994). Multisensory control of escape in the cockroach Periplaneta americana. II. Patterns of touch-evoked behavior. *Journal of Comparative Physiology A* **174(1)**, 13–26

[51] Comer, C. M.; Parks, L.; Halvorsen, M. B.; Breese-Tertelling, A. (2003). The antennal system and cockroach evasive behavior. II. Stimulus identification and localization are separable antennal functions. *Journal of Comparative Physiology A* **189**, 97–103

[52] Connell, J. H. (1990). *Minimalist Mobile Robotics: A Colony Architecture for an Artificial Creature*. Cambridge, Massachusetts: Academic

[53] Consi, T.; Grasso, F.; Mountain, D.; Atema, J. (1995). Explorations of turbulent odor plumes with an autonomous underwater robot. *Biological Bulletin* **189**, 231–232

[54] Coulter, D. (2000). *Digital Audio Processing*. Lawrence, KS: CMP Books

[55] Cruse, H. (2002). The functional sense of central oscillations in walking. *Biological Cybernetics* **86(4)**, 271–280

[56] Cruse, H.; Bläsing, B.; Dean, J.; Dürr, V.; Kindermann, T.; Schmitz, J.; Schumm, M. (2004). WalkNet—a decentralized architecture for the control of walking behaviour based on insect studies. In: F. Pfeiffer;

T. Zielinska (eds.), *Walking: Biological and Technological Aspects*, Berlin Heidelberg New York: Springer, pp. 81–118

[57] Dahl, F. (1883). Über die Hörhaare bei den Arachniden. *Zoologischer Anzeiger* **6**, 267–270. [Translation of title: On the sensory hairs in Arachnida]

[58] Delcomyn, F. (1980). Neural basis of rhythmic behavior in animals. *Science* **210**, 492–498

[59] Delcomyn, F. (2004). Insect waking and robotics. *Annual Review Entomology* **49**, 51–70

[60] Drewes, C. D.; Bernard, R. A. (1976). Electrophysiological responses of chemosensitive sensilla in the wolf spider. *Journal of Experimental Zoology* **198(3)**, 423–435

[61] Dumpert, K. (1978). Spider odor receptor: Electrophysiological proof. *Experientia* **34**, 754–755

[62] Dürr, V.; König, Y.; Kittmann, R. (2001). The antennal motor system of the stick insect Carausius morosus: Anatomy and antennal movements during walking. *Journal of Comparative Physiology A* **187(2)**, 131–144

[63] Dürr, V.; Schmitz, J.; Cruse, H. (2004). Behaviour-based modelling of hexapod locomotion: Linking biology and technical application. *Arthropod Structure and Development* **33(3)**, 237–250

[64] Ekeberg, Ö.; Blümel, M.; Büschges, A. (2004). Dynamic simulation of insect walking. *Arthropod Structure and Development* **33**, 287–300

[65] Endo, G.; Nakanishi, J.; Morimoto, J.; Cheng, G. (2005). Experimental studies of a neural oscillator for biped locomotion with QRIO. In: *Proceedings of the IEEE International Conference on Robotics and Automation (ICRA)*, pp. 596–602

[66] Espenschied, K. S.; Quinn, R. D.; Beer, R. D.; Chiel, H. J. (1996). Biologically based distributed control and local reflexes improve rough terrain locomotion in a hexapod robot. *Robotics and Autonomous Systems* **18**, 59–64

[67] Fend, M.; Bovet, S.; Pfeifer, R. (2006) On the influence of morphology of tactile sensors for behavior and control. *Robotics and Autonomous Systems* **54(8)**, 686–695

[68] Fend, M.; Bovet, S.; Yokoi, H.; Pfeifer, R. (2003). An active artificial whisker array for texture discrimination. In: *Proceedings of the IEEE/RSJ International Conference on Intelligent Robots and Systems (IROS)*, vol. *2*, pp. 1044–1049

[69] Fend, M.; Yokoi, H.; Pfeifer, R. (2003). Optimal morphology of a biologically-inspired whisker array on an obstacle-avoiding robot. In: *Proceedings of the Seventh European Conference on Artificial Life (ECAL)*, Berlin Heidelberg New York: Springer, pp. 771–780

[70] Ferrell, C. (1993). *Robust Agent Control of an Autonomous Robot with Many Sensors and Actuators*. Masters thesis, Massachusetts Institute of Technology, USA

[71] Ferrell, C. (1994). Robust and adaptive locomotion of an autonomous hexapod. In: P. Gaussier; J.-D. Nicoud (eds.), *Proceedings From Perception to Action Conference (PERAC 1994)*, Los Alamitos, CA: IEEE Computer Society, pp. 66–77

[72] Filliat, D.; Kodjabachian, J.; Meyer, J. A. (1999). Incremental evolution of neural controllers for navigation in a 6 legged robot. In: *Proceedings of the Fourth International Symposium on Artificial Life and Robotics*, Oita University Press, pp. 745–750

[73] Fischer, J. (2004). *A Modulatory Learning Rule for Neural Learning and Metalearning in Real World Robots with Many Degrees of Freedom.* Ph.D. thesis, University of Münster, Germany, Aachen: Shaker

[74] Fischer, J.; Pasemann, F.; Manoonpong, P. (2004). Neuro-controllers for walking machines—an evolutionary approach to robust behavior. In: M. Armada; P. Gonzalez de Santos (eds.), *Proceedings of the Seventh International Conference on Climbing and Walking Robots (CLAWAR'04)*, Berlin Heidelberg New York: Springer, pp. 97–102

[75] Floreano, D.; Mondada, F. (1994). Active perception, navigation, homing, and grasping: An autonomous Perspective. In: Ph. Gaussier; J.-D. Nicoud (eds.), *Proceedings of From Perception to Action Conference (PERAC 1994)*, Los Alamitos, CA: IEEE Computer Society, pp. 122–133

[76] Foelix, R. F.; Chu-Wang, I. W. (1973). Morphology of spider sensilla II. Chemoreceptors. *Tissue Cell* **5(3)**, 461–478

[77] Fogel, D. B.; Fogel, L. J. (1996). An introduction to evolutionary programming. In: *Proceedings of Evolution Artificielle*, Berlin Heidelberg New York: Springer, pp. 21–33

[78] Fogel, L. J.; Owens, A. J.; Walsh, M. J. (1966). *Artificial Intelligence Through Simulated Evolution.* New York: Wiley

[79] Franklin, R. F. (1985). The locomotion of hexapods on rough ground. In: M. Gewecke; G. Wendler (eds.), *Insect Locomotion*, Hamburg: Paul Parey, pp. 69–78

[80] Frik, M.; Guddat, M.; Karatas, M.; Losch, D. C. (1999). A novel approach to autonomous control of walking machines. In: *Proceedings of the Second International Conference on Climbing and Walking Robots (CLAWAR'99)*, Portsmouth: Professional Engineering, pp. 333–342

[81] Gaßmann, B.; Scholl, K.-U.; Berns, K. (2001). Locomotion of LAURON III in rough terrain. In: *Proceedings of the IEEE/ASME International Conference on Advanced Intelligent Mechatronics, vol. 2*, pp. 959–964

[82] Geng, T.; Porr, B.; Wörgötter, F. (2006). Fast biped walking with a reflexive neuronal controller and real-time online learning. *International Journal of Robotics Research* **25(3)**, 243–259

[83] Grayson, K. (2000). Urodele Amphibians: The Regenerative Vertebrate Exception. http://www.bio.davidson.edu/Courses/anphys/2000/Grayson/GRAYSON.HTM. Cited 6 December 2005

[84] Grillner, S. (1991). Recombination of motor pattern generators. *Current Biology* **1(4)**, 231–233

[85] Grillner, S.; Zangger, P. (1979). On the central generation of locomotion in the low spinal cat. *Experimental Brain Research* **34(2)**, 241–261

[86] Hagras, H.; Callaghan, V.; Colley, M. J. (2000). Online learning of fuzzy behaviour co-ordination for autonomous agents using genetic algorithms and real-time interaction with the environment. In: *Proceedings of the Ninth IEEE International Conference on Fuzzy Systems, vol. 2*, pp. 853–858

[87] Hansen, H. J. (1917). On the Trichobothria ("auditory hairs") in arachnida, myriopoda, and insects, with a summary of the external sensory organs in arachnida. *Entomologisk Tidskrift* **38**, 240–259

[88] Haschke, R. (2003). *Bifurcations in Discrete-Time Neural Networks – Controlling Complex Network Behaviour with Inputs.* Ph.D. thesis, Bielefeld University, Germany

[89] Heiden, U. an der (1991). Neural networks: Flexible modelling, mathematical analysis, and applications. In: F. Pasemann; H.-D. Doebner (eds.), *Neurodynamics*, World Scientific, pp. 49–95

[90] Hergenröder, R.; Barth, F. G. (1983). Vibratory signals and spider behavior: How do the sensory inputs from the eight legs interact in orientation? *Journal of Comparative Physiology A* **152**, 361–371

[91] Hilljegerdes, J.; Spenneberg, D.; Kirchner, F. (2005). The construction of the four legged prototype robot ARAMIES. In: *Proceedings of the International Conference on Climbing and Walking Robots (CLAWAR'05)*, Berlin Heidelberg New York: Springer, pp. 335–342

[92] Hirose, S.; Inoue, S.; Yoneda, K. (1990). The whisker sensor and the transmission of multiple sensor signals. *Advance Robotics* **4(2)**, 105–117

[93] Holland, J. H. (1975). *Adaptation in Natural and Artificial Systems: An Introductory Analysis with Applications to Biology, Control, and Artificial Intelligence.* Ann Arbor: The University of Michigan Press

[94] Hooper, S. L. (2000). Central pattern generators. *Current Biology* **10(5)**, R176–R179

[95] Horchler, A. D.; Reeve, R. E.; Webb, B.; Quinn, R. D. (2004). Robot phonotaxis in the wild: A biologically inspired approach to outdoor sound localization. *Advanced Robotics* **18(8)**, 801–816

[96] Hülse, M.; Pasemann, F. (2002). Dynamical neural Schmitt trigger for robot control. In: J. R. Dorronsoro (ed.), *Proceedings of the International Conference on Artificial Neural Networks (ICANN 2002)*, Berlin Heidelberg New York: Springer, *vol. 2415 of Lecture Notes in Computer Science*, pp. 783–788

[97] Hülse, M.; Wischmann, S.; Pasemann, F. (2004). Structure and function of evolved neuro-controllers for autonomous robots. *Connection Science* **16(4)**, 249–266

[98] Hülse, M.; Zahedi, K.; Pasemann, F. (2003). Representing robot-environment interactions by dynamical features of neuro-controllers. In:

Anticipatory Behavior in Adaptive Learning Systems, Berlin Heidelberg New York: Springer, *vol. 2684* of *Lecture Notes in Computer Science*, pp. 222–242

[99] Iida, F.; Pfeifer, R. (2006). Sensing through body dynamics. *Robotics and Autonomous Systems* **54(8)**, 631–640

[100] Ijspeert, A. J. (2001). A connectionist central pattern generator for the aquatic and terrestrial gaits of a simulated salamander. *Biological Cybernetics* **84(5)**, 331–348

[101] Inagaki, S.; Yuasa, H.; Suzuki, T.; Arai, T. (2006). Wave CPG model for autonomous decentralized multi-legged robot: Gait generation and walking speed control. *Robotics and Autonomous Systems* **54(2)**, 118–126

[102] Ingvast, J.; Ridderström, C.; Hardarson, F.; Wikander J. (2003). Warp1: Towards walking in rough terrain—control of walking. In: G. Muscato; D. Longo (eds.), *Proceedings of the Sixth International Conference on Climbing and Walking Robots (CLAWAR'03)*, London: Professional Engineering, pp. 197–204

[103] Jacobi, N. (1998). Running across the reality gap: Octopod locomotion evolved in a minimal simulation. In: P. Husbands; J.-A. Meyer (eds.), *Proceedings of the First European Workshop on Evolutionary Robotics (EvoRobot'98)*, Berlin Heidelberg New York: Springer, *vol. 1468* of *Lecture Notes in Computer Science*, pp. 39–58

[104] Jimenez, M. A.; Gonzalez de Santos, P. (1997). Terrain adaptive gait for walking machines. *International Journal of Robotics Research* **16(3)**, 320–339

[105] Jung, D.; Zelinsky, A. (1996). Whisker-based mobile robot navigation. In: *Proceedings of the IEEE/RSJ International Conference on Intelligent Robots and Systems (IROS)*, pp. 497–504

[106] Kaneko, M.; Kanayama, N.; Tsuji, T. (1998). Active antenna for contact sensing. *IEEE Transactions on Robotics and Automation* **14(2)**, 278–291

[107] Kato, K.; Hirose, S. (2001). Development of quadruped walking robot, TITAN-IX—mechanical design concept and application for the humanitarian demining robot. *Advanced Robotics* **15(2)**, 191–204

[108] Kauer, J. S. (2005). Salamander Locomotion. http://birg.epfl.ch/page45111.html. Cited 6 December 2005

[109] Kerscher, T.; Albiez, J.; Berns, K. (2002). Joint control of the six-legged robot AirBug driven by fluidic muscles. In: *Proceedings of the Third International Workshop on Robot Motion and Control (RoMoCo'02)*, pp. 27–32

[110] Kikuchi, F.; Ota, Y.; Hirose, S. (2003). Basic performance experiments for jumping quadruped. In: *Proceedings of the IEEE/RSJ International Conference on Intelligent Robots and Systems (IROS)*, pp. 3378–3383

[111] Kimura, H.; Akiyama, S.; Sakurama, K. (1999). Realization of dynamic walking and running of the quadruped using neural oscillator. *Autonomous Robots* **7(3)**, 247–258

[112] Kimura, H.; Fukuoka, Y. (2000). Biologically inspired dynamic walking of a quadruped robot on irregular terrain—adaptation at spinal cord and brain stem. In: *Proceedings of the First International Symposium on Adaptive Motion of Animals and Machines*, TuA-II-1. http://www.kimura.is.uec.ac.jp/amam2000/index.html. Cited 24 December 2005

[113] Kirchner, F.; Spenneberg, D.; Linnemann, R. (2002). A biologically inspired approach toward robust real-world locomotion in legged robots. In: J. Ayers; J. Davis; A. Rudolph (eds.), *Neurotechnology for Biomimetic Robots*, MIT Press, pp. 419–447

[114] Klaassen, B.; Zahedi, K.; Pasemann, F. (2004). A modular approach to construction and control of walking robots. In: *Robotik 2004, vol. 1841 of VDI-Berichte*, pp. 633–640

[115] Krichmar, J.L.; Nitz, D. A.; Gally, J. A.; Edelman, G. M. (2005) Characterizing functional hippocampal pathways in a brain-based device as it solves a spatial memory task. *Proceedings of the National Academy of Sciences of the United States of America* **102(6)**, 2111–2116

[116] Kuffler, S. W.; Nicholls, J. G.; Martin, A. R. (1984). *From Neuron to Brain 2nd edn.* Sunderlands, Massachusetts: Sinauer

[117] Kurazume, R.; Yoneda, K.; Hirose, S. (2002). Feedforward and feedback dynamic trot gait control for quadruped walking vehicle. *Autonomous Robots* **12(2)**, 157–172

[118] Leger, P. C. (2000). *Darwin2K: An Evolutionary Approach to Automated Design for Robotics*. Massachusetts: Kluwer Academic

[119] Lewis, M. A.; Fagg, A. H.; Solidum, A. (1992). Genetic programming approach to the construction of a neural network for control of a walking robot. In: *Proceedings of the IEEE International Conference on Robotics and Automation (ICRA)*, pp. 2618–2623

[120] Linder, C. R. (2005). Self-organization in a simple task of motor control based on spatial encoding. *International Society for Adaptive Behavior* **13(3)**, 189–209

[121] Lund, H. H.; Webb, B.; Hallam, J. (1998). Physical and temporal scaling considerations in a robot model of cricket calling song preference. *Artificial Life* **4(1)**, 95–107

[122] Lungarella, M.; Hafner, V. V.; Pfeifer, R.; Yokoi, H. (2002). An artificial whisker sensor for robotics. In: *Proceedings of the IEEE/RSJ International Conference on Intelligent Robots and Systems (IROS)*, pp. 2931–2936

[123] Luo, F. L.; Unbehauen, R. (1999). *Applied Neural Networks for Signal Processing*. Cambridge University Press

[124] Mahn, B. (2003). *Entwicklung von Neurokontrollern für eine holonome Roboterplatform*. Diplomarbeit, Fachhochschule Oldenburg, Germany

[125] Manoonpong, P.; Pasemann, F. (2005). Advanced mobility sensor driven-walking device 02 (AMOS-WD02). In: *Proceedings of the Third International Symposium on Adaptive Motion in Animals and*

Machines, Robot data sheet, Ilmenau: ISLE, p. R22. http://www.tu-ilmenau.de/fakmb/fileadmin/template/amam/div/AMOS-WD02.pdf. Cited 6 December 2005

[126] Manoonpong, P.; Pasemann, F. (2005). Advanced mobility sensor driven-walking device 06 (AMOS-WD06). In: *Proceedings of the Third International Symposium on Adaptive Motion in Animals and Machines*, Robot data sheet, Ilmenau: ISLE, p. R23. http://www.tu-ilmenau.de/fakmb/fileadmin/template/amam/div/AMOS-WD06.pdf. Cited 6 December 2005

[127] Manoonpong, P.; Pasemann, F.; Fischer, J. (2004). Neural processing of auditory–tactile sensor data to perform reactive behavior of walking machines. In: *Proceedings of the IEEE International Conference on Mechatronics and Robotics (MechRob'04)*, Aachen: Sascha Eysoldt, *vol. 1*, pp. 189–194

[128] Manoonpong, P.; Pasemann, F.; Fischer, J. (2005). Modular neural control for a reactive behavior of walking machines. In: *Proceedings of the Sixth IEEE Symposium on Computational Intelligence in Robotics and Automation (CIRA 2005)*, pp. 403–408

[129] Manoonpong, P.; Pasemann, F.; Fischer, J.; Roth, H. (2005). Neural processing of auditory signals and modular neural control for sound tropism of walking machines. *International Journal of Advanced Robotic Systems* **2(3)**, 223–234

[130] Manoonpong, P.; Pasemann, F.; Roth, H. (2006). A modular neuro-controller for a sensor-driven reactive behavior of biologically inspired walking machines. *International Journal of Computing* **5(3)**, 75–86

[131] Markelic, I. (2005). *Evolving a neurocontroller for a fast quadrupedal walking behavior*. Masters thesis, Institut für Computervisualistik Arbeitsgruppe Aktives Sehen, Universität Koblenz-Landau, Germany

[132] Matsuoka, K. (1985). Sustained oscillations generated by mutually inhibiting neurons with adaptation. *Biological Cybernetics* **52(6)**, 367–376

[133] Matsusaka, Y.; Kubota, S.; Tojo, T.; Furukawa, K.; Kobayashi, T. (1999). Multi-person conversation robot using multi-modal interface. In: *Proceedings of World Multiconference on Systems, Cybernetic and Informatics, vol. 7*, pp. 450–455

[134] Michelsen, A.; Popov, A. V.; Lewis, B. (1994). Physics of directional hearing in the cricket Gryllus bimaculatus. *Journal of Comparative Physiology A* **175**, 153–164

[135] Mondada, F.; Franzi, E.; Ienne, P. (1993). Mobile robot miniaturisation: A tool for investigation in control algorithms. In: *Proceedings of the Third International Symposium on Experimental Robotics*, Berlin Heidelberg New York: Springer, pp. 501–513

[136] Moravec, H. P. (1977). Towards automatic visual obstacle avoidance. In: *Proceedings of the Fifth International Joint Conference on Artificial Intelligence*, Cambridge, Massachusetts: Morgan Kaufmann, p. 584

[137] Murphey, R. K.; Zaretsky, M. D. (1972). Orientation to calling song by female crickets, *Scapsipedus marginatus (Gryllidae)*. *Journal of Experimental Biology* **56(2)**, 335–352

[138] Murray, J. C.; Erwin, H.; Wermter, S. (2004). Robotics sound-source localization and tracking using interaural time difference and cross-correlation. In: *Proceedings of NeuroBotics Workshop*, pp. 89–97

[139] Nafis, G. (2005). California Reptiles and Amphibians. http://www.californiaherps.com. Cited 6 December 2005

[140] Nakada, K.; Asai, T.; Amemiya, Y. (2003). An analog neural oscillator circuit for locomotion control in quadruped walking robot. In: *Proceedings of the International Joint Conference on Neural Networks, vol. 2*, pp. 983–988

[141] Nehmzow, U.; Walker, K. (2005). Quantitative description of robot-environment interaction using chaos theory. *Robotics and Autonomous Systems* **53(3–4)**, 177–193

[142] Hecht-Nielsen, R. (1990). *Neurocomputing*. Reading, Massachusetts: Addison-Wesley

[143] Nilsson, N. J. (1969). A mobile automaton: An application of artificial intelligence techniques. In: *Proceedings of the First International Joint Conference on Artificial Intelligence*, pp. 509–520

[144] Nolfi, S.; Floreano, D. (1998). Co-evolving predator and prey robots: Do 'arms races' arise in artificial evolution? *Artificial Life* **4(4)**, 311–335

[145] Nolfi, S.; Floreano, D. (2000). *Evolutionary Robotics – The Biology, Intelligence, and Technology of Self-organizing Machines*. Cambridge, Massachusetts: MIT Press

[146] Okada, M.; Nakamura, D.; Nakamura, Y. (2003). Hierarchical design of dynamics based information processing system for humanoid motion generation. In: *Proceedings of the Second International Symposium on Adaptive Motion of Animals and Machines*, SaP-III-1. http://www.kimura.is.uec.ac.jp/amam2003/ABSTRACTS/N12-okada.pdf. Cited 10 November 2005

[147] Olive, C. W. (1982). Sex pheromones in two orbweaving spiders (*Araneae, Araneidae*): An experimental field study. *Journal of Arachnology* **10**, 241–245

[148] Orlovsky, G. N.; Deliagina, T. G.; Grillner, S. (1999). *Neural Control of Locomotion. From Mollusc to Man*. New York: Oxford University Press

[149] Parker, G. B.; Lee, Z. (2003). Evolving neural networks for hexapod leg controllers. In: *Proceedings of the IEEE/RSJ International Conference on Intelligent Robots and Systems (IROS), vol. 2*, pp. 1376–1381

[150] Pasemann, F. (1993). Discrete dynamics of two neuron networks. *Open Systems and Information Dynamics* **2**, 49–66

[151] Pasemann, F. (1993). Dynamics of a single model neuron. *International Journal of Bifurcation and Chaos* **3(2)**, 271–278

[152] Pasemann, F. (1998). Structure and dynamics of recurrent neuromodules. *Theory in Biosciences* **117**, 1–17

[153] Pasemann, F. (2002). Complex dynamics and the structure of small neural networks. *Network: Computation in Neural Systems* **13**, 195–216

[154] Pasemann, F.; Hild, M.; Zahedi, K. (2003). SO(2)–networks as neural oscillators. In: J. Mira; J. R. Alvarez (eds.), *Computational Methods in Neural Modeling: Proceedings of the Seventh International Work-Conference on Artificial and Natural Neural Networks (IWANN 2003)*, Berlin Heidelberg New York: Springer, *vol. 2686 of Lecture Notes in Computer Science*, pp. 144–151

[155] Pasemann, F.; Hülse, M.; Zahedi, K. (2003). Evolved neurodynamics for robot control. In: M. Verleysen (ed.), *European Symposium on Artificial Neural Networks*, pp. 439–444

[156] Pasemann, F.; Steinmetz, U.; Hülse, M.; Lasa, B. (2001). Robot control and the evolution of modular neurodynamics. *Theory in Biosciences* **120**, 311–326

[157] Payton, D. W. (1986). An architecture for reflexive autonomous vehicle control. In: *Proceedings of the IEEE International Conference on Robotics and Automation (ICRA)*, pp. 1838–1845

[158] Pearson, K. G. (1976). The control of walking. *Scientific American* **235(6)**, 72–86

[159] Pearson, K. G.; Iles, J. F. (1973). Nervous mechanisms underlying intersegmental co-ordination of leg movements during walking in the cockroach. *Journal of Experimental Biology* **58**, 725–744

[160] Petersen, S. R. (2005). BugWeb. http://www.bugweb.dk. Cited 2 December 2005

[161] Pfeifer, R.; Scheier, C. (1999). *Understanding Intelligence*. Cambridge, Massachusetts: MIT Press

[162] Pfeiffer, F.; Weidemann, H. J.; Eltze, J. (1994). The TUM walking machine. *Intelligent Automation and Soft Computing* **2**, TSI, 167–174

[163] Pfeiffer, F.; Zielinska, T. (eds.) (2004). *Walking: Biological and Technological Aspects*. Berlin Heidelberg New York: Springer

[164] Quinn, R. D.; Nelson, G. M.; Bachmann, R. J.; Kingsley, D. A.; Offi, J.; Ritzmann, R. E. (2001). Insect designs for improved robot mobility. In: K. Berns; R. Dillmann (eds.), *Proceedings of the Fourth International Conference on Climbing and Walking Robots Conference (CLAWAR'01)*, London: Professional Engineering, pp. 69–76

[165] Radinsky, L. B. (1986). *The Evolution of Vertebrate Design*. Chicago: Chicago University Press

[166] Randall, M. J. (2001). *Adaptive Neural Control of Walking Robots*. London: Professional Engineering

[167] Rechenberg, I. (1973). *Evolutionsstrategie—Optimierung technischer Systeme nach Prinzipien der biologischen Evolution*. Stuttgart: Fommann-Holzboog

[168] Reeve, R. (1999). *Generating Walking Behaviors in Legged Robots*. Ph.D. thesis, University of Edinburgh, UK

[169] Reeve, R.; Webb, B. (2002). New neural circuits for robot phonotaxis. *Philosophical Transactions of the Royal Society A: Mathematical, Physical and Engineering Sciences* **361(1811)**, 2245–2266

[170] Renals, S.; Rohwer, R. (1990). A study of network dynamics. *Journal of Statistical Physics* **58**, 825–848

[171] Righetti, L.; Ijspeert, A. J. (2006). Programmable central pattern generators: An application to biped locomotion control. In: *Proceedings of the IEEE International Conference on Robotics and Automation (ICRA)*, pp. 1585–1590

[172] Ritzmann, R. E. (1984). The cockroach escape response. *Neural Mechanisms of Startle Behavior*, New York, London: Plenum, 93–131

[173] Ritzmann, R. E.; Quinn, R. D.; Fischer, M. S. (2004). Convergent evolution and locomotion through complex terrain by insects, vertebrates and robots. *Arthropod Structure and Development* **33(3)**, 361–379

[174] Rojas, R. (1996). *Neural Networks—A Systematic Introduction*. Berlin Heidelberg New York: Springer

[175] Roth, H.; Schilling, K. (1996). Hierarchically organised control strategies for mobile robots based on fused sensor data. In: *Proceedings of Mechatronics, vol. 1*, pp. 95–100

[176] Roth, H.; Schwarte, R.; Ruangpayoongsak, N.; Kuhle, J.; Albrecht, M.; Grothof, M.; Heß, H. (2003). 3D Vision based on PMD-technology and fuzzy logic control for mobile robots. In: *Proceedings of the Second International Conference on Soft Computing, Computing with Words and Perceptions in System Analysis, Decision and Control (ICSCCW 2003)*, pp. 9–11

[177] Roth, H.; Schwarte, R.; Ruangpayoongsak, N.; Kuhle, J.; Albrecht, M.; Grothof, M.; Heß, H. (2003). 3D Vision based on PMD-technology for mobile robots. In: *Proceedings of the SPIE, Aerosense—Technologies and Systems for Defense and Security Conference, vol. 5083*, pp. 556–567

[178] Rupprecht, K. U. (2004). *Entwicklung eines Gelenkelementes einer Laufmaschine*. Diplomarbeit, Fachbereich Maschinenbau, Rheinische Fachhochschule Köln, Germany

[179] Russell, A.; Thiel, D.; Mackay-Sim, A. (1994). Sensing odour trails for mobile robot navigation. In: *Proceedings of the IEEE International Conference on Robotics and Automation (ICRA)*, pp. 2672–2677

[180] Saranli, U.; Buehler, M.; Koditschek, D. E. (2001). RHex: A simple and highly mobile hexapod robot. *International Journal of Robotics Research* **20(7)**, 616–631

[181] Schaefer, P. L.; Kondagunta, G. V.; Ritzmann, R. E. (1994). Motion analysis of escape movements evoked by tactile stimulation in the cockroach Periplaneta americana. *Journal of Experimental Biology* **190**, 287–294

[182] Schmidhuber, J.; Zhumatiy, V.; Gagliolo, M. (2004). Bias-optimal incremental learning of control sequences for virtual robots. In: *Proceedings*

of the Conference on Intelligent Autonomous Systems (IAS-8), Amsterdam: IOS, pp. 658–665

[183] Schneider, D. (1964). Insect antennae. *Annual Review of Entomology* **9**, 103–122

[184] Schneider, D. (1999). Insect pheromone research: Some history and 45 years of personal recollections. *Insect Semiochemicals-IOBC, West Palearctic Regional Section (WPRS), Bulletin* **22(9)**. http://phero.net/iobc/dachau/bulletin99/schneider.pdf. Cited 5 December 2005

[185] Schwefel, H. P. (1995). *Evolution and Optimum Seeking*. New York: Wiley.

[186] Seyfarth, E. A. (1985). Spider proprioception: Receptors, reflexes, and control of locomotion. In: F. G. Barth (ed.), *Neurobiology of Arachnids*, Berlin Heidelberg New York: Springer, pp. 230–248

[187] Seyfarth, E. A. (2002). Tactile body raising: Neuronal correlates of a 'simple' behavior in spiders. In: *Proceedings of the European Colloquium of Arachnology, European Arachnology 2000*, Aarhus: Aarhus University Press, pp. 19–32

[188] Shik, M. L.; Severin, F. V.; Orlovskii, G. N. (1966). Control of walking and running by means of electrical stimulation of the mid-brain. *Biophysics* **11**, 756–765

[189] Smith, R. (2004). *Open Dynamics Engine v0.5 User Guide*. http://www.ode.org/ode-0.5-userguide.html. Cited 18 December 2005

[190] Song, S. M.; Waldron, K. J. (1989). *Machines That Walk: The Adaptive Suspension Vehicle*. Cambridge, Massachusetts: MIT Press

[191] Spenneberg, D.; Kirchner, F. (2002). SCORPION: A biomimetic walking robot. In: *Robotik 2002, vol. 1679* of *VDI-Berichte*, pp. 677–682

[192] Spenneberg, D.; McCullough, K.; Kirchner, F. (2004). Stability of walking in a multilegged robot suffering leg loss. In: *Proceedings of the IEEE International Conference on Robotics and Automation (ICRA), vol. 3*, pp. 2159–2164

[193] Stein, P. S. G; Grillner, S.; Selverston, A. I.; Stuart, D. G. (eds.) (1997). *Neurons, Networks, and Behavior*. Cambridge, Massachusetts: MIT Press

[194] Stierle, I.; Getman, M.; Comer, C. M. (1994). Multisensory control of escape in the cockroach Periplaneta americana I. Initial evidence from patterns of wind-evoked behavior. *Journal of Comparative Physiology A* **174**, 1–11

[195] Svaizer, P.; Matassoni, M.; Omologo, M. (1997). Acoustic source location in a three-dimensional space using cross-power spectrum phase. In: *Proceedings of the IEEE International Conference on Acoustics, Speech, and Signal Processing (ICASSP'97), vol. 1*, pp. 231–234

[196] Svarer, C. (1994). *Neural Networks for Signal Processing*. Ph.D. thesis, Electronics Institute, Technical University of Denmark, Denmark

[197] Taga, G.; Yamaguchi, Y.; Shimizu, H. (1991). Self-organized control of bipedal locomotion by neural oscillators in unpredictable environment. *Biological Cybernetics* **65(3)**, 147–159

[198] Takemura, H.; Deguchi, M.; Ueda, J.; Matsumoto, Y.; Ogasawara, T. (2005). Slip-adaptive walk of quadruped robot. *Robotics and Autonomous Systems* **53(2)**, 124–141

[199] Todd, D. J. (1985). *Walking Machines: An Introduction to Legged Robots*. London: Kogan Page

[200] Tomasinelli, F. (2002). Isopoda. http://www.isopoda.net. Cited 4 December 2005

[201] Toyomasu, M.; Shinohara, A. (2003). Developing dynamic gaits for four legged robots. In: *Proceedings of International Symposium on Information Science and Electrical Engineering*, pp. 577–580

[202] Twickel, A. (2004). *Obstacle perception by scorpions and robots*. Masters thesis, University of Bonn, Germany

[203] Twickel, A.; Pasemann, F. (2005). Evolved neural reflex-oscillators for walking machines. In: J. Mira; J. R. Alvarez (eds.), *Proceedings of the First International Work-Conference on the Interplay between Natural and Artificial Computation (IWINAC 2005)*, Berlin Heidelberg New York: Springer, *vol. 3561 of Lecture Notes in Computer Science*, pp. 376–385

[204] Twickel, A.; Pasemann, F. (2006). Reflex-oscillations in evolved single leg neurocontrollers for walking machines. *Natural Computing*, Berlin Heidelberg New York: Springer, DOI: http://dx.doi.org/10.1007/s11047-006-9011-y

[205] Uncini, A. (2003). Audio signal processing by neural networks. *Neurocomputing* **55(3-4)**, 593–625

[206] Valin, J. M.; Michaud, F.; Rouat, J.; Létourneau, D. (2003). Robust sound source localization using a microphone array on a mobile robot. In: *Proceedings of the IEEE/RSJ International Conference on Intelligent Robots and Systems (IROS)*, pp. 1228–1233

[207] Wadden, T. (1988). *Neural Control of Locomotion in Biological and Robotic Systems*. Ph.D. thesis, Royal Institute of Technology, Sweden

[208] Walter, W. G. (1953). *The Living Brain*. New York: Norton

[209] Wang, Q. H.; Ivanov, T.; Aarabi, P. (2004). Acoustic robot navigation using distributed microphone arrays. *Information Fusion (Special Issue on Robust Speech Processing)* **5(2)**, 131–140

[210] Watanabe, K.; Hashem, M. M. A. (2004). *Evolutionary Computations: New Algorithms and Their Applications to Evolutionary Robots*. Berlin Heidelberg New York: Springer

[211] Watson, J. T.; Ritzmann, R. E.; Zill, S. N.; Pollack, A. J. (2002). Control of obstacle climbing in the cockroach, *Blaberus discoidalis* I. Kinematics. *Journal of Comparative Physiology A* **188**, 39–53

[212] Webb, B. (1998). Robots crickets and ants: Models of neural control of chemotaxis and phonotaxis. *Neural Networks* **11**, 1479–1496

[213] Webb, B. (2000). What does robotics offer animal behaviour? *Animal Behaviour* **60(5)**, 545–558

[214] Webb, B. (2001). Can robots make good models of biological behaviour? *Target Article for Behavioural and Brain Sciences* **24(6)**, 1033–1094

[215] Webb, B.; Scutt, T. (2000). A simple latency dependent spiking neuron model of cricket phonotaxis. *Biological Cybernetics* **82(3)**, 247–269

[216] Wei, T. E.; Quinn, R. D.; Ritzmann, R. E. (2004). A CLAWAR that benefits from abstracted cockroach locomotion principles. In: M. Armada; P. Gonzalez de Santos (eds.), *Proceedings of the Seventh International Conference on Climbing and Walking Robots (CLAWAR'04)*, Berlin Heidelberg New York: Springer, pp. 849–858

[217] Wischmann, S.; Hülse, M.; Pasemann, F. (2005). (Co)evolution of (de)centralized neural control for a gravitationally driven machine. In: *Proceedings of the European Conference on Artificial Life (ECAL 2005)*, Berlin Heidelberg New York: Springer, pp. 179–188

[218] Wischmann, S.; Pasemann, F. (2004). From passive to active dynamic 3D bipedal walking—An evolutionary approach. In: M. Armada; P. Gonzalez de Santos (eds.), *Proceedings of the Seventh International Conference on Climbing and Walking Robots (CLAWAR'04)*, Berlin Heidelberg New York: Springer, pp. 737–744

[219] Wongsuwan, H.; Laowattana, D. (2004). Bipedal gait synthesizer using adaptive neuro-fuzzy network. In: *Proceedings of the First Asia International Symposium on Mechatronics (AISM)*, pp. 433–438

[220] Wörgötter, F.; Porr, B. (2005) Temporal sequence learning, prediction and control—A review of different models and their relation to biological mechanisms. *Neural Computation* **17(2)**, 245–319

[221] Yamaguchi, T.; Watanabe, K.; Izumi, K. (2005). Neural network approach to acquiring free-gait motion of quadruped robots in obstacle avoidance. *Artificial Life and Robotics* **9(4)**, 188–193

[222] Yokoi, H.; Fend, M.; Pfeifer, R. (2004). Development of a whisker sensor system and simulation of active whisking for agent navigation. In: *Proceedings of the IEEE/RSJ International Conference on Intelligent Robots and Systems (IROS)*, pp. 607–612

[223] Yoneda, K.; Ota, Y.; Ito, F.; Hirose, S. (2001). Quadruped walking robot with reduced degrees of freedom. *Journal of Robotics and Mechatronics* **13(2)**, 190–197

[224] Zahedi, K.; Hülse, M.; Pasemann, F. (2004). Evolving neurocontrollers in the RoboCup domain. In: *Robotik 2004*, vol. *1841* of *VDI-Berichte*, pp. 63–70

[225] Zaknich, A. (2003). *Neural Networks for Intelligent Signal Processing. vol. 4 of series on Advanced Biology and Logic-Based Intelligence*, Singapore: World Scientific

[226] Zhang, Y.; Weng, J. (2001). Grounded auditory development by a developmental robot. In: *Proceedings of the INNS/IEEE International Joint Conference on Neural Networks*, pp. 1059–1064

Index

Activation function, 34, 36
Additive neuron model, 34
Advanced auditory network, 72–74, 76,
    82, 86
Agent–agent interactions, 150
Agent–environment interactions, 2, 8
ANNs, *see* Artificial neural networks
Antenna-like sensors, 53, 147
Artificial
    auditory–tactile sensor, 113
    hairs, 7
    insect, 4
    neuron, 35
    whisker sensor, 48
Artificial neural networks, 31, 36
Artificial perception–action system,
    10–11
Attractors, 39, 77
Auditory signal processing, 67–68, 80,
    116, 148
Auditory signal processing network, 135
Auditory-tactile sensor, 48–50, 67, 82,
    115, 117, 147, 148
Auditory-tactile signal processing
    network, 86, 116
Autonomous Intelligent System, 149
Autonomous mobile robots, 1, 48
Axon, 32

Behavior control, 9, 10, 99
Behavior fusion, 142
Behavior fusion controller, 107, 110,
    126, 142, 148
Bias, 34, 38, 92

Biological neuron, 32–34
Biologically inspired
    locomotion control, 29
    walking machines, 147
Braitenberg vehicles, 3

Central pattern generator, 26, 91
Chemical trails, 5
Chemotaxis, 5
Cockroach, 17–18, 24
Coexisting fixed point, 39
Composite mode, 108, 142
CPG, *see* Central pattern generator
Cricket, 6, 17
    phonotaxis, 6, 14
    song, 6, 50
Cusp catastrophe, 39
Cutoff frequency, 69, 72, 76

Deliberate control, 9
Driven actions, 106
Dynamical neural Schmitt Trigger, 68

Edge following, 5
Embodied systems, 2, 149
ENS[3], *see* Evolution of neural systems
    by stochastic synthesis
Even loop, 79, 88, 126
Evolution of neural systems by
    stochastic synthesis, 41
Evolutionary
    algorithm, 41, 75, 76
    process, 44, 74
Evosun, 43

Excitatory, 33
    self-connection, 37–39, 41, 73, 77, 84, 88, 125
Exploratory behavior, 7

Fast Fourier transform, 82, 120
Feedforward network, 37
Feeding food, 5
FFT, *see* Fast Fourier transform
Finite state machines, 4
Firing rate, 33
Fitness function, 74, 77, 83
Fixed-action pattern, 14
Fusion technique, 105

Gedanken experiments, 2

Hidden neurons, 36
Hinton, 43
Hysteresis effect, 37, 39, 40, 69–70, 77, 88–90, 102
Hysteresis loop, 40, 41, 80

Infrared sensor, 52
Inhibitory, 33
    self-connection, 37–39
Input neurons, 36
Insect antennas, 15, 17
Integrated structure evolution environment, 43, 60, 68
ISEE, *see* Integrated structure evolution environment

Khepera robot, 6, 87

Layers, 37
Locomotion control, 10, 91
Look-up table, 105
Low-pass filter, 69, 70, 75, 115, 148

MBoard, *see* Multi-Servo IO-Board
Mean squared error, 74, 77
Mechanoreceptor, 17
MERLIN, 8
Mesothoracic, 25
Metathoracic, 25
Microphone, 48, 50
Minimal recurrent controller, 76, 87
Modular neural controller, 98
Modular neural structure, 9, 148

MRC, *see* Minimal recurrent controller
Multi-Servo IO-Board, 152, 153
Multiplication-like function, 94

Negative tropism, 14, 104, 149
    avoid obstacles, 2, 10
    escape behavior, 10, 14
    obstacle avoidance behavior, 6, 8, 10, 14–18, 55, 87, 128, 131
Neimark–Sacker bifurcation, 91
Neural
    control, 9, 98, 148
    preprocessing, 9, 67, 68, 75, 98, 148
Neural oscillator network, 89, 148

Obstacle avoidance controller, 102, 148
Obstacle avoidance mode, 108, 142
Operator
    evaluation, 43
    mutation, 43
    reproduction, 42
    selection, 43
Orthopteran escape response, 6
Output neurons, 37

Periplaneta computatrix, 5
Photonic Mixer Device (PMD), 8
Physical sensor systems, 47
Population, 41
Positive tropism, 14, 104, 149
    light stimulus, 2
    photo, 6
    sound tropism, 11, 22, 48, 49, 75, 79, 136–142, 149
Potentiometer, 48
Predator escape behavior, 18
Preprocessing of antenna-like sensor data, 67, 123, 148
Prey capture behavior, 14, 20, 103
Proprioceptors, 149
Prothoracic, 25

Quasi-periodic attractors, 91

Reactive
    behavior, 14, 104, 142, 147, 149
    control, 9
    system, 9
    walking machines, 151

Recurrent
    network, 37
    neural network, 6, 10
    neuro-module, 37
Reflexes, 14
Rhythmic patterned activity, 26–27
RNNs, see Recurrent neural network,
    Recurrent network
Robot behavior, 6
Robotic
    behavior, 13
    systems, 2

Salamander, 23, 24
Scorpions, 15
    Leiurus quinquestriatus, 15
    Pandinus cavimanus, 15, 47
Seeking food, 5, 15
Self-connection, 37, 38, 69, 77, 79, 87
Sensor fusion technique, 11, 126
Sensorimotor loop, 10
Servomotor modules, 154
Signal processing network of antenna-
    like sensors, 89, 99
Simple auditory network, 68, 70
Soma, 32
Sound tropism controller, 103, 135
Sound–direction detection network,
    77–80, 120, 123
Spider, 19–20
    Cupiennius salei, 19, 47, 49
Spinal cord, 26
Stable fixed point, 39
Stance phase, 25, 28
Stereo auditory sensor, 49, 52, 67, 75,
    76, 147
Subsumption architecture, 4
Swing phase, 25, 28
Synapses, 32
Synaptic strength, 35

Tactile sensing systems, 14

Tactile signal processing, 67, 83, 86
Taxes, 14
TDOA, see Time delay of arrival
Threshold value, 72, 76, 139
Time delay of arrival, 50, 76, 121
Time scheduling, 105
Touch-evoked behavior, 17
Transfer function, 35, 36
    hyperbolic, 36
    linear, 35
    linear threshold, 35
    sigmoid, 35
Trichobothria, 19, 20
Tripod gait, 94
Trot gait, 94

Unstable fixed point, 39
Update frequency, 44, 68, 76, 87, 118

Velocity regulating network, 89, 94
Versatile artificial perception–action
    systems, 11, 37, 51, 104, 149
VRN, see Velocity regulating network

Walking animals, 22
Walking gaits, 27
Walking machines, 10, 48, 53, 55
    four-legged, 50, 53, 142, 148
        AMOS-WD02, 53, 57–60, 76, 99,
            120, 131, 151
    six-legged, 4, 129, 148
        AMOS-WD06, 53, 61–63, 87, 99,
            134, 152
        Genghis, 4
    Whegs, 6
Walter's tortoise, 2
Wandering, 5, 15, 143
Wind-mediated escape, 7

YARS, see Yet another robot simulator
Yet another robot simulator, 60

Cognitive Technologies

J. W. Lloyd:
Logic for Learning.
Learning Comprehensive Theories from Structured Data.
X, 256 pages. 2003

S. K. Pal, L. Polkowski, A. Skowron (Eds.):
Rough-Neural Computing.
Techniques for Computing with Words.
XXV, 734 pages. 2004

H. Prendinger, M. Ishizuka (Eds.):
Life-Like Characters.
Tools, Affective Functions, and Applications.
IX, 477 pages. 2004

H. Helbig:
**Knowledge Representation and
the Semantics of Natural Language.**
XVIII, 646 pages. 2006

P. M. Nugues:
**An Introduction to Language Processing
with Perl and Prolog.**
An Outline of Theories, Implementation,
and Application with Special Consideration
of English, French, and German.
XX, 513 pages. 2006

W. Wahlster (Ed.):
SmartKom: Foundations of Multimodal Dialogue Systems.
XVIII, 644 pages. 2006

B. Goertzel, C. Pennachin (Eds.):
Artificial General Intelligence.
XVI, 509 pages. 2007

O. Stock, M. Zancanaro (Eds.):
PEACH – Intelligent Interfaces for Museum Visits.
XVIII, 316 pages. 2007

V. Torra, Y. Narukawa:
**Modeling Decisions: Information Fusion
and Aggregation Operators**
XIV, 284 pages. 2007

P. Manoonpong:
**Neural Preprocessing and Control
of Reactive Walking Machines.**
Towards Versatile Artificial Perception–Action Systems.
XVI, 185 pages. 2007